Book 1

Sciencewise

Discovering Scientific Process through Problem Solving

written by
Dennis Holley

illustrated by
Kate Simon Huntley

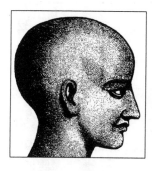

© 1996
CRITICAL THINKING BOOKS & SOFTWARE
www.criticalthinking.com
P.O. Box 448 • Pacific Grove • CA 93950-0448
Phone 800-458-4849 • FAX 831-393-3277
ISBN 0-89455-647-9
Printed in the United States of America

Table of Contents

Creative Challenges

This book is dedicated to
my wife, for her patience and understanding;
my parents, for encouraging me to wonder and explore;
my students, for teaching me more than they will ever know.

Introduction

The most beautiful thing we can experience is the mysterious. It is the source of all true art and science.

—Albert Einstein

There seem to be two aspects to the uniquely human enterprise we call science. One aspect is the continual search for natural truths. The other aspect is the vast matrix of facts and knowledge. Unfortunately, many parents and students, most textbook publishers, and even some teachers have the mistaken notion that science is memorizing terms, names, and facts. Nothing could be further from the truth.

Science is built up with facts, as a house is with stones. But a collection of facts is no more a science than a heap of stones is a house.

—Jules Henri Poincare

Science cannot be gleaned from the glossy pages of a pricey textbook. Science must be experienced, not memorized. Facts and knowledge are really the accumulated fruits of centuries of scientific labor. Make no mistake, humans are able to use these facts and this knowledge in the most incredible and creative ways imaginable, but facts alone are not science.

Imagination is more important than knowledge.

—Albert Einstein

The true essence of science is the relentless and unwavering need to know "why." It is this need, this nagging, irresistible curiosity to search for answers, that drives our species from the depths of the ocean to the blackness of outer space. This is what science really is, and kids of any gender, color, and practically any age, with the proper guidance, can <u>do</u> real science.

Unfortunately, not only do students come to you not fully understanding science, they also lack the skills necessary to actually <u>do</u> science. They do, however, come through your classroom door fully charged with the most important precursor to scientific inquiry—curiosity. With this remarkable inquisitiveness, students, with your help, can begin to learn the skills necessary to <u>do</u> science. There are five basic science skills students need to develop: observing, predicting, designing/experimenting, eliminating, and drawing conclusions.

Observing

Science done right requires students to be accurate and thorough observers. Learning good observation techniques requires much practice. Hand in hand with the development of these skills must come the realization that any observation is the unique perspective of the observer, colored and altered by his/her experiences, expectations, and emotions. Students must constantly be challenged to separate inference (what they think is there) from reality (what is actually there) in their observations.

In the field of observation, chance favors the prepared mind.

—Louis Pasteur

Predicting

Curiosity raises questions. Careful observation reveals information. Using this information, we make predictions about the possible answers to our questions.

Prediction is probably the easiest skill

for students to master. The main problem you will encounter is that students are often reluctant to make predictions for fear of being wrong. This is an attitude you must constantly strive to change. In science, right or wrong predictions don't matter. Science is the search for natural truths, and it matters not whether you come in the front door (correct predictions) or the back door (incorrect predictions) of the house of truth. What matters is that either way, in the end, you learn the truth.

Along with the development of this skill and this attitude must come the realization that certain predictions are more valid than others. A hypothesis is merely a guess, conjecture, or untested speculation. A theory is a higher level of prediction because it is an educated guess based on some evidence or past experience.

Designing/Experimenting

With the realization that speculation and prediction must be tested in controlled experiments to determine the truth, modern science was born. For too long, people accepted the musings of authority figures as truth and fact. Often the more bizarre the speculation, the more eager people were (and some still are) to believe it. Experimentation is what separates science from philosophy and superstition.

> *The practice of science enables scientists as ordinary people to go about doing generally ordinary things that, when assembled, reveal the extraordinary intricacies and awesome beauties of nature.*
> —Arthur Kornberg

The first hurdle to clear in experimental design is to determine what problem is to be solved. Problems should always be stated in question form, be as simple and specific as possible, and address only one factor at a time.

Let us use a whimsical imaginary problem to demonstrate experimental design. Suppose you had to solve the following problem: What is the effect on aardvarks of eating chocolate pudding? Assuming you had unlimited resources, your experimental design might go something like this:

Step 1
Determine what problem is to be solved. The problem—What is the effect on aardvarks of eating chocolate pudding?—is in question form, is specific, and deals with only one problem, so we are ready to proceed.

Step 2
Get 1,000 aardvarks. The more experimental subjects you work with, the more reliable is your data.

Step 3
Separate the aardvarks into two equal groups. In each group, put 500 aardvarks with the same age, sex, and physical characteristics. You want the two groups to be as nearly equal in all respects as possible.

Step 4
Keep both groups under identical conditions—same size cages, same amount and kind of food, same amount of water, same period of light and dark, same temperature, same humidity, and so on. These conditions, called the control variables, must be kept as nearly identical as possible.

Step 5
Feed one group of aardvarks chocolate pudding, and designate it the experimental group. Do not feed chocolate pudding to the other group, and designate it the control group. The chocolate

pudding in the experimental group is called the *manipulated variable*; it is the effect of this pudding that you are testing.

Step 6
Let the experiment run for a reasonable length of time. Collect appropriate data; for example, after two weeks the experimental group turns green and starts to do back flips.

Step 7
To verify these results, repeat the experiment as many times as possible with different aardvarks and different batches of pudding.

Step 8
Based on the accumulated data, draw reasonable conclusions. The reasonable conclusion here would be that apparently chocolate pudding causes aardvarks to turn green and do back flips.

Some problems are difficult or impossible to test experimentally. For example, to calculate the orbit of a comet, we would deduce certain outcomes then look to nature for verification.

The only solid piece of scientific truth about which I feel totally confident is that we are profoundly ignorant about nature.
—Lewis Thomas

Eliminating
Not only will you have to battle students' fear of making wrong predictions, you will also have to deal with students' fear of failure. Students must come to understand and believe that failure in science is not to be feared. Actually, we learn more from failure than success because failure raises more questions than success, and these questions, in turn, force even more inquiry. Science is dynamic and always changing. Newly discovered "facts" and even those laws of science that have withstood the test of time must be subject to revision at any moment. What we regard as facts are at best momentary illusions seen through a veil of ignorance. Today we laugh at the idea that earlier people thought it factual that the earth was flat or that living things could arise spontaneously from dead or inorganic matter. The future will show many of our so-called facts to be just as wrong.

It is possible that every law of nature so far has been incorrectly stated.
—J.B.S. Haldane

Students need to learn what data is appropriate to collect and how to organize data. Charting and graphing skills are essential. Organizing data into tables, charts, and graphs enables us to view the results in a graphic format. In this form, data is easier to understand and patterns are more easily discerned. Students must learn to deal with the fact that data may be "muddy"—inconsistent, unexpected, and often unfathomable.

Drawing Conclusions
What does the data mean? This is often difficult for professional scientists to answer, let alone students. "Muddy" data can only

yield "cloudy" conclusions, and students must learn to deal with this frustrating problem. Only experience will allow students to identify and support those conclusions that are valid and discard those that are not valid.

*The art of becoming wise is the
art of knowing what to overlook.*
—William James

Once students have some proficiency in the above skills, they can begin to think scientifically and actually <u>do</u> science using the method illustrated below:

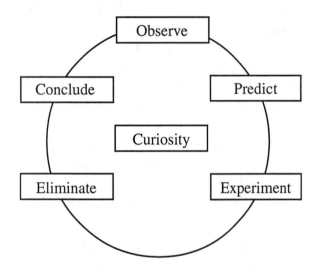

The scientific process is cyclic and becomes progressively more refined in view of the problem you are trying to solve.

Scientists approach problems in an organized way, but hunches and serendipity also play a role. What do Velcro, penicillin, x-rays, and Teflon have in common? Serendipity!—the lucky accident of finding valuable things not specifically sought after.

Science can be tedious, frustrating, and just plain boring with repeated failure often the norm. It can also be the most challenging, stimulating, and rewarding thing a person

can do. The only way for students to learn what science is and how it works is to let students actually <u>do</u> science.

*Tell me, I forget
Show me, I remember
Involve me, I understand*
—Chinese proverb

Dynamo Demos

The teacher-led Dynamo Demos (pp. 1–130) help students to develop the science process skills: observing, predicting, experimenting, eliminating data, and drawing conclusions. In addition, students develop their problem solving and creative/critical thinking skills.

In the Dynamo Demo activities, students do the thinking and the teacher does the doing. The teacher sets up and presents a "What will happen if...?" or "Why did that happen?" problem situation. Using guided questions and the necessary manipulation of apparatus and equipment, the teacher helps the students understand the problem, make accurate observations and reasonable predictions, and arrive at a conclusion or an answer to the problem.

While specific scientific principles and concepts are demonstrated in the Dynamo Demos activities, the primary focus is on actively involving students in the scientific process and developing problem solving and creative/critical thinking skills.

When possible, the Dynamo Demos are counterintuitive and present discrepant results. This approach makes the Demos more challenging and often results in incorrect predictions, forcing students to "come in the back door" to the truth.

The first demonstration is a preliminary activity designed to refine students' observation skills. The rest of the demonstrations can be done in any order.

> *We must not confuse the results of science with the ways in which scientists produce those results. The practice of science is a social process, an art or a craft and not a science.*
> —Michael Klapper

Creative Challenges

The student-centered Creative Challenges (pp. 131–71) help students develop their creative/critical-thinking, problem-solving, and "inventioneering" skills.

In the Dynamo Demos, the teacher sets up an experiment and uses select questions to guide the students to a solution of the given problem. With the Creative Challenges, the teacher presents the problem (challenge) then functions merely as a technical advisor. Stu-dents must design and develop a solution to the problem (challenge). This allows students to practice and apply the science process skills demonstrated in the Dynamo Demos.

The student combines a process of "solution generation" with trial and error to look for possible solutions to a given problem. Each time a possible solution is generated, the student must predict what will happen and then observe the results when the solution is implemented. Each solution that is tried requires analysis and interpretation of the results and allows the student to draw conclusions on how to best solve the original problem. Specific scientific principles and concepts are learned through the process itself rather than through direct teacher instruction.

> *Invention is a combination of brains and material. The more brains you use, the less material you need.*
> —Charles F. Kettering

On Using This Book

The following are general information, guidelines, and hints for using this book:

- Use this book in conjunction with any commercial or teacher-authored science curriculum.

- The activities are designed to come from "left field." That is, the activities are not intended to teach any concept being discussed in class. The less the Demos and Challenges have to do with what is going on in class, the better. Students are then forced to think and create rather than turn to a book for the answer.

- The activities are not written to be formally graded. Assess and/or grade them as you see fit. I, personally, do not grade my students on the Demos or Challenges as I am not sure how you accurately evaluate and assess creativity and critical thinking. Furthermore, I worry that formal grading will stifle creativity and make these activities just another educational burden for students to endure.

- These activities produce creative effort and critical thinking, not marketable results. Failure and incorrect predictions are expected and even encouraged. Such is the reality of actual everyday science.

- Students should work individually as much as possible. If time and/or materials present a problem, students may work in design teams of two, preferably, but three if necessary. The larger the design teams, the less each team member will get from the activity and the more conflicts you will have within groups.

- Many of the challenges have a competitive element. Students need to be competitive and learn how to deal with competition; however, students should not lose sight of the fact that competition is secondary in these activities. Creative thinking and effort are the primary goals. Stress the process, not the product.

- Encourage students to do their own thinking and creating. Don't let competitive fervor lead to outside input by parents, siblings, and/or peers. Along those same lines, set spending limits on those challenges where students are required to supply their own construction materials.

- Try the activities before you present them to your class. The activities in this book are tried and tested, and I have attempted to write the directions in clear and concise terms. But play it safe, work things out beforehand. Don't waste precious class time by "winging it."

- Once Demos and Challenges are complete, encourage students, when practical, to take this learning out of the classroom and share it with parents, siblings, and friends. This is what I call the "Ripple Effect." Educational research shows that we best learn what we teach to others. Always consider the safety, practicality, and the necessity of supervision when having students do these activities outside the classroom.

- Give students recognition for their efforts. This might include displaying their creations and inventions, awarding and displaying appropriate certificates of achievement, and/or presenting gag gifts as trophies. Such recognition will generate a great deal of student enthusiasm, parental support, and positive public relations.

- Kids may clamor to do these activities every day, but that is not the intent of this book. This book is designed to supplement an established curriculum. You should do one or two Demos first before starting students on the individual Challenges. There are enough Dynamo Demos in this book to do one demo every week and enough Creative Challenges to do one every other week during a normal-length school year. You certainly can do fewer than this, but in my experience, it may not be practical to do more from a time standpoint.

- Ask, don't tell. It is crucial that you help students develop their thinking skills by asking for student observations, predictions, and explanations rather than just giving the answer. Mix closed questions that demand factual recall with open questions that are divergent and thought-provoking. Use questions to generate discussion, but don't always call on the student who raises his hand, and give adequate wait time when you question. Space does not permit a detailed discussion here of how to construct good oral questions, but there are excellent resources available to help you develop sound questioning strategies.

In closing, be creative yourself with this book. Take my offerings and put your own twist to them. If you come up with a better way of doing some of the things in this book or with whole new activities that could be included in future editions of this book, contact me.

Dennis Holley
Critical Thinking Books & Software
P.O. Box 448
Pacific Grove, California 93950

Dynamo Demos

1 A New "Friend"

Problem: *How observant are you?*

Observe

1. Accurate observation plays a very important role in science. Unfortunately, people are not born with this skill; observation can only be developed with training and practice. How good is your observation skill? In this activity, you will be challenged to develop and practice your powers of observation.

2. You will work individually on this activity.

3. The teacher will give you a new "friend." In the space below, make as accurate an observation as possible of this object in the time you are given. HINT: The quickest and most accurate way to record your observation is through a combination of diagrams (pictures) and written description, so both draw and describe your new "friend." Do not, however, write on or mark your "friend" in any way.

4. Once everyone has completed the observation, the teacher will pick up your "friend" and give you one more challenge to meet.

---

1   For the Teacher

Objective

In this activity, students will use a peanut to hone their observation skills.

Materials Needed

- peanuts in the shell, one peanut per student

Curiosity Hook

Sit and shell a few peanuts as students come into the classroom, but keep the peanut container hidder.

Setup

1. Start this activity by telling students you want to introduce them to a new "friend." Once you have tweaked your students' curiosity, give each student one peanut.

2. Challenge each student to accurately observe and describe his/her peanut "friend." Space has been provided on the student page for the description. Encourage students to use use both diagrams and written description.

3. Give students a reasonable length of time to complete their observations and descriptions then pick up all the peanuts. Put all the students' peanuts back into the original container. Question the class about the accuracy and completeness of their observations. Are they confident they were accurate and complete?

4. Pour all the peanuts out of the container into a large pile. Mix the pile up. Now, challenge the students to find the exact peanut they just observed and described. Are they still as confident about their observations? It will be less noisy and confusing to have only a few students at a time search for their peanut.

Safety Concerns

Use common sense.

Outcomes and Explanations

Discuss with students the type of detailed information that makes for a good observation. What were some of the variables students used to identify their peanuts? Discuss with students why accurate observation is a key part of the scientific process.

Application

Inventors are creative and observant people who see connections, patterns, and solutions that others often overlook. Newton hypothesized the laws of gravitation while observing an apple fall.

Flickering Flames

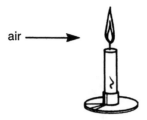

Problem: *Can you predict which way the flame will flicker?*

Predict

1. Predict what will happen when the teacher blows directly towards a candle flame.

Conclude

2. Which way did the flame flicker in demonstration 1? Why did it happen that way?

Predict

3. Predict what will happen when the teacher blows towards a candle flame from behind a note card.

Conclude

4. What effect did the note card have on the direction of the flame? Why?

Predict

5. Predict what will happen when the teacher blows towards a candle flame from behind a jar.

Conclude

6. In demonstration 3, did the jar have the same effect as the note card on the direction of the flame? Why or why not?

2 For the Teacher

Objective

In this activity, students will use their powers of observation and critical thinking to determine which way a candle flame will flicker.

Materials Needed

- 1 candle
- 1 note card, 3" × 5"
- 1 small glass or plastic jar
  (preferably painted in bright colors)

Curiosity Hook

Have the candle set up and lit when the students come in. Place the candle in a highly visible but safe location and, if possible, place a brightly colored jar near the candle.

Setup

1. Place the candle where stray air movements will not affect the flame; you need an undisturbed flame to begin each demonstration. The larger the candle the better, as small candles are difficult for students to see. Turning off the classroom lights might also help students to see better the flame and how it moves.

   Have students record their predictions as they observe an undisturbed flame. Now, blow with a gentle but sustained flow of air. Position yourself so the airflow is directed squarely at the candle flame.

2. Place a note card between the source of airflow and the candle flame. Have students record their predictions as they observe an undisturbed candle flame with the note card being held about 4 to 8 inches away from the candle flame.

   Hold the note card about 2 to 4 inches from your lips and position the card about 4 to 8 inches away from the candle flame. Hold the bottom of the note card between the thumb and forefinger of one hand. This prevents your hand from interfering with the flow of air. Again, blow towards the candle with a gentle but sustained flow of air. Because of the card, a slightly more vigorous flow of air will be needed to initiate an effect.

3. Place a jar between the source of airflow and the candle flame. Have students record their predictions as they observe an undisturbed candle flame with a jar placed about 3 to 4 inches away from the candle. Any round container will work, but a quart jar or a two-liter plastic pop bottle is ideal. Gallon jars will work but necessitate a much more vigorous flow of air to initiate an effect.

   As before, blow with a gentle but sustained flow of air. I use a jar painted with wild patterns in bright colors. I tell my students that it is a magic jar and that I can blow right through it. I then conclude demonstration 3 by blowing one very hard burst of air directly at the jar in line with the candle flame, extinguishing the flame.

Safety Concerns

Use common sense when handling an open flame.

Outcomes and Explanations

1. A candle flame will bend away from a direct flow of air.

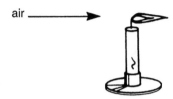

Have students write the explanation under 2 on the student pages.

2. The candle flame will bend towards the note card. The note card causes air turbulence, thus reducing the air pressure behind the card. The candle flame is pushed towards the card by the greater air pressure around the candle flame. Discuss with students the effect of the note card on the candle flame and have students write the explanation under 4 on the student pages.

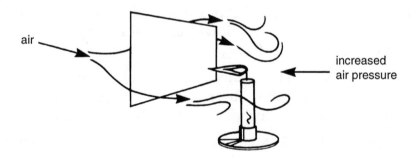

3. The candle flame will bend away from the air flowing around the jar. The air flows around the jar and onto the flame. A forceful burst of air will extinguish the flame because the jar directs the airflow at the flame. Discuss with students the effect the jar has on the flame and have them write the explanation under 6 on the student pages.

Application

Have students apply what they have learned by imagining the following scenario: You (the student) are the famous automobile engineer, Neil Wheel. You have been assigned the task of designing the rear end of a new car. The car must move through the air with as little disturbance of the air as possible. Will you design the car with a square rear end or a rounded rear end and why?

Take Home

Encourage your students to demonstrate the magic jar in demonstration 3 to their parents, siblings, or friends. Caution them to do so only under adult supervision and remind them of the potential hazards of something so apparently innocent as a candle flame.

3　Funny Funnel

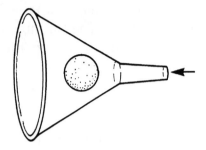

Problem:　*Duck! The teacher is going to try and shoot Ping-Pong balls out of a funnel.*

Predict

1. Predict what will happen when the teacher blows through a funnel containing a Ping-Pong ball.

Conclude

2. What did the Ping-Pong ball do? Why did it happen that way?

Predict

3. Predict what will happen when the teacher points the funnel and the ball down and blows through it.

Conclude

4. What did the Ping-Pong ball do this time? Why?

Predict

5. Based on your previous conclusions, how can you meet the teacher's challenge?

Conclude

6. How was the teacher able to lift the Ping-Pong ball? Why?

3 | For the Teacher

Objective

In this activity, a discrepancy involving a Ping-Pong ball and a funnel will test students' powers of observation, prediction, critical thinking, and problem solving.

Materials Needed

- 1 Ping-Pong ball

- 1 glass funnel
  (A plastic funnel will work, but unless it is clear plastic, it will hide the ball from view and lessen the effect of the activity.)

Curiosity Hook

Have the funnel and Ping-Pong ball out where students can see them as they come into the classroom. I hold the funnel over the ball and swirl the ball around inside the funnel as students come in.

Setup

Have students observe, describe, and predict what will happen to the Ping-Pong ball in each demonstration below. Encourage students to use both labeled diagrams and written descriptions.

1. Hold the Ping-Pong ball in the funnel with your finger. Remove your finger and let the ball roll out of the funnel several times. Now, point the bell of the funnel at a nearby student, blow hard into the funnel, and remove your finger from the ball while you are blowing. The student may duck and cringe, but the ball will not come out of the funnel.

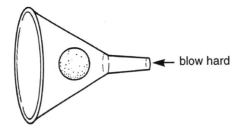
← blow hard

2. Again, hold the Ping-Pong ball in the funnel with your finger. Point the funnel down towards the floor, blow hard, and remove your finger while you are blowing. The ball will again "stick" in the funnel.

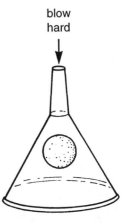
blow hard

3. Place the Ping-Pong ball on a table and challenge the students to think of a way to pick up the ball using only the funnel. They are not allowed to directly touch the ball.

Safety Concerns

For sanitary reasons, demonstrate student suggestions yourself. If students try them, clean the neck of the funnel each time with soap and water or alcohol and rinse thoroughly.

Outcomes and Explanations

1. As you blow into the funnel, the airflow removes some of the air behind the ball. This lowers the air pressure behind the ball. The greater air pressure in front of the ball pushes the ball back into the funnel.

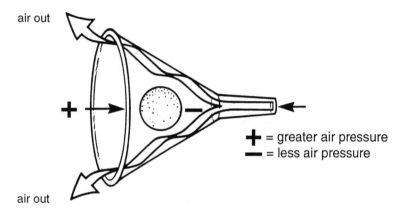

air out

air out

+ = greater air pressure
— = less air pressure

Discuss this result with students and have them write the explanation under 2 on the student pages.

2. The explanation is the same for the second demonstration. In fact, the harder you blow, the tighter the ball is jammed into the funnel. Have students write the explanation under 4 on the student pages.

3. To meet the challenge of lifting the ball, place the funnel on the table over the ball. Blow hard and continue to do so as you slowly lift the funnel. The ball should "stick" in the funnel and come up off the table as the funnel is lifted. Discuss this effect with students and have them write the explanation under 6 on the student pages.

Application

This activity is a classic example of Bernoulli's principle, which states: The higher the speed of a flowing liquid or gas, the lower the pressure. Blowing air into the funnel lowers the air pressure behind the ball, so the greater pressure outside pushes the ball into the funnel. You can see this effect in tennis and baseball. Top spin on a tennis ball creates a stronger and faster airflow under the ball and curves the ball's path downward. A spin throw of a baseball makes the air rush by faster on the side of the ball which is spinning forward, reducing the pressure on that side and curving the ball's path in the direction of the spin.

Take Home

If students have access to a Ping-Pong ball and funnel, encourage them to demonstrate the funny funnel to their parents, siblings, or friends. Urge students to use caution if working with glass funnels. Broken glass always presents a health threat. Also, instruct students not to put small objects in the funnel. Very small objects might be aspirated up through the funnel into a student's mouth and cause choking.

4 Weightless Water

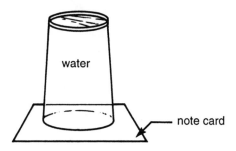

water

note card

Problem: *Has the teacher really discovered weightless water?*

Predict

1. Predict what will happen when the teacher places a note card over a glass of water, turns the glass upside down, and then removes his/her hand.

Conclude

2. What happened when the teacher removed his/her hand and why did it happen?

Predict

3. Predict what you think will happen when the teacher places other objects over a glass of water and then turns the glass upside down.

Conclude

4. What effect did the other objects have when the glass was turned upside down and why?

Predict

5. Predict what you think will happen when the teacher uses liquids other than water.

Conclude

6. What effect did other liquids have on the experiment?

 © 1996 Critical Thinking Books & Software • P.O. Box 448, Pacific Grove, CA 93950 • 800-458-4849

4   For the Teacher

Objective

In this activity, gravity seems to be nullified. Students use their powers of observation and critical thinking to understand what keeps water from falling out of a cup when the cup is turned upside down.

Materials Needed

- 1 clear glass or plastic cup
- several note cards
- a piece of paper, a piece of cardboard from a box, a piece of Masonite, and a piece of wood, all the same approximate dimensions as the note card
- 1 container of rubbing (isopropyl) alcohol
  (or enough to fill the glass)
- 1 container of carbonated drink
  (or enough to fill the glass)

Curiosity Hook

Have a glass of water with a note card on it in view when students come into the classroom.

Setup

1. Once students have made their predictions under 1 on the student page, fill a glass to overflowing then slide a note card across the top of the glass so that it cuts off the water bulging above the top of the glass. Hold the note card in place with your hand, turn the glass upside down, and slowly remove your hand. The note card and the water should remain in place.

2. Try other materials like paper, cardboard, Masonite, and wood instead of the note card. Have students write their predictions for each material under 3 on the student pages.

3. Try other liquids, like alcohol and carbonated beverages, instead of water. Have students write their predictions for each type of liquid under 5 on the student pages.

Safety Concerns

This can get messy, so I suggest you do each setup over a sink and have plenty of towels handy. Alcohol can cause serious gastric disturbances if taken internally. Dispose of alcohol by pouring it on a sidewalk and letting it evaporate.

Outcomes and Explanations

1. Many students will predict that the water will fall out of a glass that is covered with a note card and turned upside down.

2. Why doesn't the water fall out? There are two forces working on the water— (1) gravity pulling down on the water, the note card, and the glass and (2) air pressure pushing up on the note card. In this case, the upward force (air pressure) is greater than the downward force (gravity).

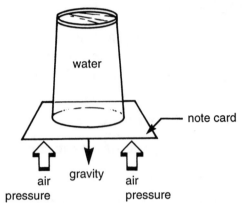

Discuss with students the effects of gravity and air pressure, and have them write the explanation under 2 on the student pages.

3. Other materials will work as well as a note card. Paper and cardboard from a box will work, but Masonite or wood will not. Why? The paper and cardboard are light, but the Masonite and wood are heavy enough that the downward force of gravity now becomes greater in the equation, thus overcoming the upward force of air pressure and pulling down the water and the Masonite or wood.

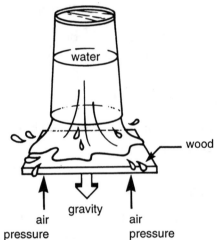

Discuss with students the differences in materials used, and have them write the explanation under 4 on the student pages.

4. Other liquids will work instead of water. Alcohol will work, but the carbonated beverage will not. Why? The carbon dioxide bubbles from the beverage exert pressure in the glass. That pressure, along with the downward pull of gravity, becomes greater in the equation, thus overcoming the upward force of air pressure and pulling down the carbonated beverage and the note card.

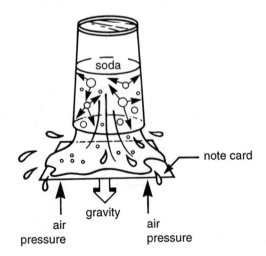

Discuss this difference in liquids with students and have them write the explanation under 6 on the student pages.

Application

How much air pressure is exerted on the note card? How much force is actually preventing the water from falling out of the glass? Challenge students to make the necessary calculations:

Step 1—Measure the dimensions of the note card.

Step 2—Multiply the length of the card by the width of the card to get the area of the card in square inches.

Step 3—Multiply the area of the card by 14.7 pounds per square inch of air pressure. Atmospheric pressure at sea level is 14.7 pounds per square inch, and since air pressure decreases as you go up, it is probably less where you are (a lot less if you are on a mountain top).

For example:

Step 1—A note card measures 3" × 5".

Step 2—The surface area of the card is 3 × 5 = 15 square inches.

Step 3—Multiply the 15 square inches × 14.7 pounds per square inch = 220.5 pounds of air pressure on the note card. The discrepancy between air appearing as nothingness and the fact that it really does exert tremendous pressure will astound students.

Take Home

This activity can be easily and safely done by students outside the classroom. However, this activity can get messy, so caution them to conduct their experiments in a location where spilling water will not be a problem.

5 Hungry Jar

Problem: *Can you get the balloon into the jar?*

Predict

1. In the space below, list all the ways you can think of to get a water-filled balloon into a jar.

Observe

2. The teacher will now show you one possible way to solve this problem. Describe what the teacher did to get the balloon into the jar.

Conclude

3. Why did it work?

Conclude

4. The teacher demonstrated one technique for getting the balloon into a jar. Can we apply the same principle to get the balloon out of the jar? List as many ways as you can think of to get the balloon out of the jar.

5 For the Teacher

Objective

In this activity, it appears that a water-filled balloon is sucked into a jar. Students will need to use their powers of observation and critical thinking and draw on their personal experiences to understand that things are not always as they seem.

Materials Needed

- 1 clear glass jar or container
  (Do not use plastic because you will be burning paper in the container. The size of the container is not critical but the larger the container, the more dramatic and visible the effect.)

- 1 balloon

- water to fill the balloon

- a paper towel or small piece of newspaper

- matches

Curiosity Hook

Place a water-filled balloon on the mouth of a jar and place this setup where students can see it as they enter the classroom. Catch student attention by using the brightest- or wildest-colored balloon possible.

Setup

1. Fill the balloon with water until it is about twice as big around as the mouth of the jar, and place the balloon on the mouth of the jar.

2. Challenge students to think of ways to get the balloon into the jar. Have them write their solutions under 1 on the student page. The problem presented has no rules or restrictions, so accept any reasonable answer.

3. To show the students your method of getting the balloon into the jar, remove the balloon, crumple up a paper towel or small piece of newspaper, and light the paper on fire. Drop the flaming paper into the jar. The paper should burn for several seconds. As the flames die down or just as they go out, place the balloon back on the mouth of the jar. The balloon will appear to be sucked down into the jar.

4. Now, challenge students to list all the ways they can think of to get the balloon out of the bottle. Have them list their ideas under 4 of the student page.

Safety Concerns

1. Because you will be working with burning paper, wear safety glasses.

2. To prevent a burn, use tongs to hold the burning paper prior to dropping it into the jar.

3. I've never had a jar break while doing this demo, but to be on the safe side, do not let students crowd up close to the jar.

Outcomes and Explanations

1. How does the burning paper method work? The burning paper uses up some of the air in the jar, and the smoke coming out of the jar carries some of the air in the jar out with it. Both of these factors result in less air inside the jar than outside the jar.

air out

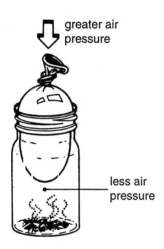

greater air pressure

less air pressure

When you quickly seal the jar with the water-filled balloon, a pressure difference results, and the greater pressure outside pushes the balloon into the jar. Write the explanation under 3 on the student page.

2. How do you get the balloon out of the jar? There are no rules, so accept any reasonable answer. To apply the same principle, rinse out the ashes, turn the jar over, and blow hard on the opening (or you may want to use a hair dryer) to force air around the balloon and into the jar. When you stop blowing, the increased air pressure above the balloon should push the balloon out of the jar.

Application

Objects are not sucked into a vacuum cleaner; objects are pushed into a vacuum cleaner. Students have a hard time accepting this fact because their eyes seem to tell them otherwise. The motor of the vacuum cleaner pulls air out of the bag or canister of the vacuum cleaner. This lowers the air pressure in the bag or canister, and the greater air pressure outside the bag or canister pushes air in. As the air rushes in, it carries with it the dog hair, cracker crumbs, dust, or whatever else you are trying to pick up.

Take Home

Caution students attempting your burning paper method at home to do so only under adult supervision. A serious fire could result if this demo is not conducted using good safety procedures.

6 | Boil That Balloon

Problem: *Will a balloon pop when heated?*

Predict

1. What will happen when the teacher heats a balloon filled only with air?

Conclude

2. What did happen when the teacher heated a balloon filled only with air, and why did it happen that way?

Predict

3. What will happen when the teacher heats a balloon filled with water?

Conclude

4. What did happen when the teacher heated a balloon filled with water?

Conclude

5. Explain why it happened.

6 For the Teacher

Objective

Put a balloon over an open flame, but do not have it pop. Sound impossible? Students will think so until they use their powers of observation, critical thinking, and problem solving to discover that, with the application of the proper scientific principles, even the seemingly impossible becomes feasible.

Materials Needed

- 2 balloons
  (Size doesn't matter but the larger the balloon, the more dramatic the effect will be and, thus, easier for students to see.)

- 1 ring-stand setup
  (If necessary, borrow this from your school's chemistry department.)

- 1 heat source
  (You need an open flame, so a hot plate will not work. Use either a Bunsen burner or a large candle.)

- safety goggles

- tongs to hold the balloon filled with air

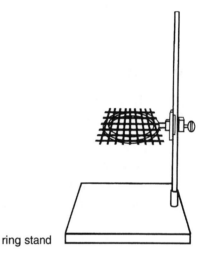

ring stand

Curiosity Hook

Have a beaker of colored liquid (food coloring in water) boiling on the ring stand when students come into the classroom.

Setup

1. Light the candle or Bunsen burner. If you are using a Bunsen burner, adjust the air supply until you get a nice flickering yellow flame. This type of flame won't be as hot as a blue flame, and it will show up more dramatically.

2. Blow up a balloon and tie it off. Ask students to predict what will happen if you hold the balloon over the flame. Have students write their predictions under 1 on the student page. The air-filled balloon will pop almost immediately.

3. Now fill a balloon about 1/2 to 3/4 full of water and tie it off. Ask students to predict what will happen if you place this balloon over the flame. Have students write their predictions under 3 on the student page.

4. Place the water-filled balloon on the screen of the ring stand and move the heat source under it. The flames will lick up around the balloon, but it will not burst.

Safety Concerns

1. You should wear safety goggles while conducting this demonstration.

2. Position students a safe distance from the heat source and the balloons because when the air-filled balloon pops, it could send small shards of hot rubber flying in all directions.

3. You will be working around an open flame, so be careful to keep clothes and hair away.

4. Position the ring stand so that it is not sitting on or near anything flammable.

5. Do not leave the water-filled balloon over the flame too long. As the water inside heats, the balloon will start to swell and could burst, spraying boiling water all over.

Outcomes and Explanations

1. The air-filled balloon pops almost immediately because the heat quickly raises the temperature of the skin of the balloon to the melting point, releasing the air inside. Have students write this explanation under 2 on the student page.

2. However, not only will the water-filled balloon not burst, but it will remain intact even when the water inside reaches the boiling point. How can this be? The secret is heat transferral. The heat of the flame is transferred across the skin of the balloon and into the water. Thus, the skin of the balloon never gets hot enough to melt. Have students write this explanation under 5 on the student page.

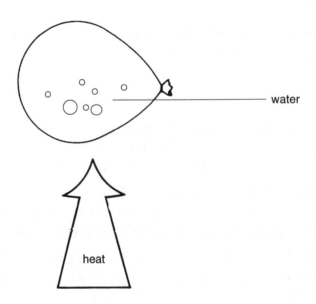

Application

Have students apply what they have learned by imagining the following scenario: You (the student) and your little cousin wish to use some old newspaper to help start a fire on your camping trip. However, you discover that the paper got wet, and now you can't get the paper to burn. What is the problem?

ANSWER: The water in the paper is absorbing the heat from the matches (or whatever you are using to try and light it), so the paper never gets hot enough to catch on fire, and it never will until it is perfectly dry.

Take Home

Students should not attempt this activity at home. Exploding balloons of air or boiling water, plus the necessity of an open flame, present potential safety hazards.

7   Leaks and Dribbles

Problem: *Are all leaks the same?*

Predict

1. On the diagram below, draw what you think will happen if the teacher pokes three holes in a container of water.

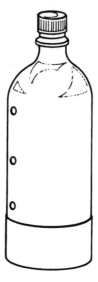

Conclude

2. What did happen when the teacher made three holes in the container? Why did it happen that way?

7   For the Teacher

Objective

In this activity, students will use their powers of observation, prediction, critical thinking, and problem solving to learn that all leaks are not created equal.

Materials Needed

- 1 plastic two-liter pop container
- a compass, awl, or anything sharp and pointed to poke a round hole in the plastic container
- a sink, bucket, or dishpan

Curiosity Hook

Have a pop container full of colored liquid (food coloring in water) leaking from a single hole. Set this up where students can see it as they come into the classroom.

Setup

1. Fill a two-liter plastic pop container with colored water and screw the lid on. Have students hypothesize what will happen if you poke three holes in the container. Have students draw their predictions on the diagram under 1 on the student page.

2. Lay the container on its side, and poke three holes in it similar to the pattern shown below.

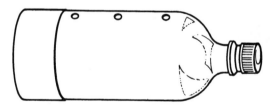

3. Stand the bottle upright, remove the lid, and let the water spray into a sink, bucket or dishpan.

Safety Concerns

Be cautious when poking holes in the container.

Outcomes and Explanations

Water will spray the greatest distance with the most force from the bottom hole, and more water will spray through the middle hole than the top hole. Why? The deeper you go in water, the greater the pressure (weight) of water pushing down on you. The bottom hole has more water pushing down on it than the other two holes. Therefore, the water pressure is least at the top hole, greater at the middle hole, and greatest at the bottom hole. Thus, the water is pushed with more pressure (force) out of the bottom hole than out of the other two holes. Discuss the effect of increasing water pressure on the holes, and have students write the explanation for the demonstration under 2 on the student page.

Application

Draw the diagrams below on a chalkboard or overhead, and challenge students to explain which dam has the greatest pressure at the base and why.

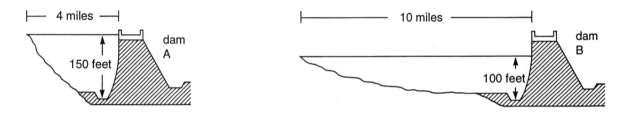

Answer: Dam A would have the greatest water pressure on its base. It is the depth, not the amount of water behind the dam, that determines pressure.

Take Home

Encourage your students to demonstrate and explain this activity to their parents, siblings, or friends.

Strange Straws

Problem: *Let's see who will win the strange-straws drinking race.*

Predict

1. Which student will win the one-straw drinking race?

Conclude

2. Which student won and why?

Predict

3. Two students will again have a drinking race. This time, one student will have two straws. The student with two straws must have one straw inside and one straw outside the glass. Which student do you think will win the drinking race?

Conclude

4. Which student won the race? What effect did the second straw have on the outcome?

8 | For the Teacher

Objective

In this activity, one student is always doomed to lose the race. Students will use their powers of observation and critical thinking to discover why the person with the strange straw can never win the drinking race.

Materials Needed

- 6 plastic drinking straws
  (Any size will work.)

- a needle to poke holes in 1 straw

- 4 glasses

- enough fruit juice to fill 4 glasses
  (You could use soda pop for this, but fruit juice would certainly set a better example healthwise.)

Curiosity Hook

Have a clear glass of colored liquid (food coloring in water) with straws in it sitting where students see it as they come into the classroom.

Setup

1. Before students come in, use a small needle to poke holes (8–10) the entire length of one straw. This is the strange straw in this situation. Do not tell students what you have done to this straw.

2. Fill two glasses with fruit juice. Put the strange straw in one glass and a regular straw in the other glass. Pick two students as contestants, and have them race to see who can drink the most juice in a set period of time. Have the class predict which contestant will win the race. Have them write their predictions under 1 on the student page. After the race, compare the amount of juice remaining in each glass and declare a winner. You can reveal the secret of the strange straw as you explain the outcome of the contest.

3. Again, fill two glasses with fruit juice and pick two students as contestants. Give one contestant a single regular straw, but give the other contestant two regular straws. The contestant with two straws must drink with both straws in the mouth but only one straw in the juice. The other straw must be outside the glass. The straw outside the glass is the strange straw in this situation.

4. Have the class predict which contestant will win the race. Have them write their predictions under 3 on the student page. Have the contestants race to see who can drink the most juice in a set period of time.

5. Compare the amount of juice remaining in each glass and declare a winner. Most students will probably have predicted that the student with two straws would win.

Safety Concerns

Caution your contestants to drink rapidly but sanely. Wild gulping could lead to a choking emergency.

Outcomes and Explanations

1. You are able to drink through a straw because of differences in air pressure. As you expand your lungs (suck on the straw), the air pressure in your mouth drops lower than the air pressure on the surface of the liquid. The greater air pressure on the liquid then pushes the liquid up the straw and into your mouth.

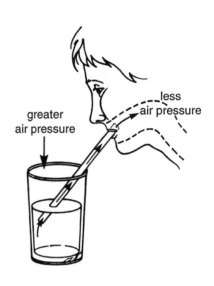

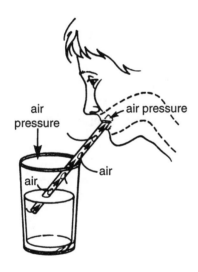

2. The contestant with the strange straw can never win this race. Air coming in through the straw with holes or through the straw outside the glass equalizes the air pressure (or nearly so) between the inside of the mouth and the top of the liquid. With no pressure difference, there is no push on the liquid and nothing moves up the straw. Have students write the explanation under 2 and 4 on the student pages.

Application

Present this question to students: If straws were built into the helmets of space suits, would astronauts on the surface of the moon be able to drink liquids through the straw?

ANSWER: No. There is no air on the moon. No air means no air pressure. Without air pressure, there would be no force to push the liquid up the straw.

Take Home

This is an easy activity to try at home.

9 Dancing Spheres

Problem: *What shape is a sphere?*

Observe

1. Describe below each of the three contraptions your teacher presents to you.

Predict

2. Predict what you think will happen to the small spheres when the teacher blows gently between them.

Conclude

3. What did happen and why?

Predict

4. Predict what will happen when the teacher gently blows between two larger spheres.

Conclude

5. Did the larger and heavier spheres behave the same way as the smaller and lighter spheres?

Predict

6. Does the shape of the objects make any difference? Will this experiment work with objects other than spheres? The teacher will do the same thing with two objects that are not spheres. Predict what will happen when the teacher gently blows between two objects that are not spheres.

Conclude

7. What did happen? Did the objects that were not spheres behave the same way as the spheres? Why or why not?

9   For the Teacher

Objective

In this activity, objects of different shapes and sizes suspended on strings will challenge students to observe, predict, reason, and understand.

Materials Needed

You will need three sets of suspended objects. I suggest you have all three sets ready to go before the activity begins. To do so, you will need the following:

- 2 small spheres
  (Ping-Pong balls will work, but I prefer golf balls. Ping-Pong balls are so lightweight they blow around all over the place; the heavier golf balls are less susceptible to air movement and give a better effect.)

- 2 larger spheres
  (Tennis balls, apples, or oranges will work.)

- 2 objects that are not spheres
  (I use square building blocks from my wife's daycare center.)

- 6 pieces of light string or heavy thread, each about 8 to 12 inches long

- 6 pieces of tape

Curiosity Hook

Have only the small-sphere setup in view when students enter the classroom. I also give each sphere a little push so they are swinging when students come in.

Setup

1. Tape one end of a piece of string to each object.

2. Tie each string to a horizontal support so that each set of objects hangs down about 6 to 8 inches from the support.

3. In each setup, each pair of objects should be hung at the same height and about 1 to 1.5 inches apart.

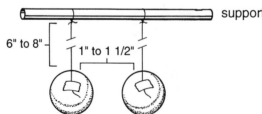

4. Have students observe and describe each setup one at a time. I recommend you have all three setups ready to go, but only show the students the small sphere setup

at first. Bring out the other setups one at a time and have students describe them as you work through the activity on the student pages. Student descriptions of each setup should go under 1 on the student page, so instruct students at the beginning of the activity to leave enough room to describe three different setups.

5. Position yourself about 4 to 5 inches away so you can blow a short, forceful burst of air directly between the objects. Practice ahead of time to determine how close you should be to the objects. The larger and heavier the objects, the more forceful the burst of air that is required to initiate an effect.

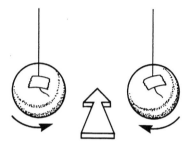

Safety Concerns

Use common sense.

Outcomes and Explanations

1. Most students will predict that the spheres will move apart, when in reality they come together when you blow between them.

2. Why do they come together? As you blow between the objects, you move the air, thus lowering the air pressure between the objects. The greater pressure on the outside of the objects pushes them together.

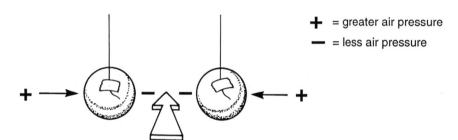

+ = greater air pressure
– = less air pressure

Discuss the effect of decreasing and increasing air pressure on the objects, and have students write the explanation for the first demonstration under 3 on the student pages.

3. The effect of blowing will be the same regardless of the size or shape of the objects (up to a certain point, so don't use bowling balls). Have students write the explanations for the next two demonstrations under 5 and 7 of the student pages.

Application

This activity demonstrates Bernoulli's principle. See Demo 3 and Demo 28 for further discussion of this effect.

10   Faulty Funnel

Problem: *Which funnel system will work?*

Observe

1. In the box to the left, use diagrams and labels to describe Funnel System A.

Funnel System A

Predict

2. What will happen when the teacher pours water in Funnel System A?

Conclude

3. What did happen when the teacher poured water in Funnel A and why did it happen that way?

Observe

4. In the box to the left, use diagrams and labels to describe Funnel System B.

Funnel System B

Predict

5. What will happen when the teacher pours water in Funnel System B?

Conclude

6. What happened when the teacher poured water in Funnel B and why did it happen that way?

7. Which funnel has a problem, and how could you fix the problem?

10 For the Teacher

Objective

In this activity, students will use their powers of observation, prediction, critical thinking, and problem solving to determine why a funnel refuses to take water.

Materials Needed

- one- and two-hole rubber stoppers

- 2 glass flasks or jars

- 2 funnels
  (If you do not have stoppers and flasks, the funnel may be inserted loosely in a jar for System A and clay can be used in place of a stopper for System B. Be sure that the opening around the funnel in System B is sealed tightly.)

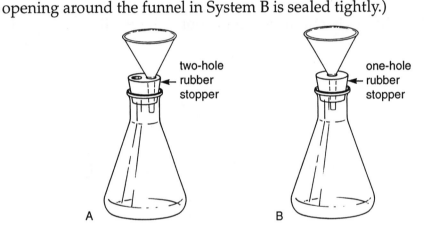

- colored liquid (food coloring in water) to pour into the funnel systems

Curiosity Hook

Have the labeled funnel systems in view when students enter the classroom.

Setup

1. It is important that the stoppers fit tightly in the jars/flasks. This demo will not work (water will flow into both funnels) if the funnels are not sealed tightly.

2. Have students predict what will happen when you try to pour water into Funnel System A. Have them write their predictions under 2 on the student pages.

3. Now pour the water into Funnel System A. The water should flow down the funnel freely.

4. Have students predict what will happen when you try to pour water into Funnel System B. Have them write their predictions under 5 on the student pages.

5. Now attempt to pour water into Funnel System B. The water should flow partway down the funnel and then stop.

Safety Concerns

Use common sense.

Outcomes and Explanations

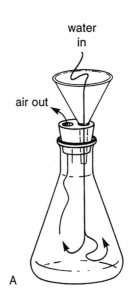

1. The water flows freely into Funnel System A because as the water comes down the funnel and into the jar/flask, it pushes the air out the other hole of the rubber stopper. The key is that the air can flow freely in and out. Have students write the explanation under 3 on the student page.

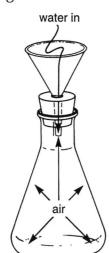

2. The water will not flow into Funnel System B because as the water comes down the funnel, it compresses (squeezes) the air in the jar/flask. Since the air in the jar can't go anywhere, the air pushes back on the water, preventing the water from moving down the funnel. Have students write this explanation under 6.

3. How can we get water into Funnel System B? Somehow we must allow the air to escape as the water enters the funnel. Accept any reasonable method(s) students come up with to solve the problem and test student methods if feasible.

Application

Students may have seen this phenomenon in action if they have ever tried to fill a narrow-mouthed bottle with water by holding it under a large stream of water from the tap. Since there is only one opening for the water to go in and air to come out, the air blocks the liquid from going in.

Take Home

Students should be able to do this activity outside of class using the loose funnel and clay seal method.

11 Bottle with a Hole

Problem: *A bottle with a hole in it will leak when filled with water. Right?*

Predict

1. The teacher has a plastic pop bottle with a hole in it. Predict what will happen when the bottle is filled with water but the lid is not put on.

Conclude

2. What did happen, and why did it happen that way?

Predict

3. Predict what will happen when the bottle is filled with water and the lid is put on.

Conclude

4. Did putting the lid on the bottle have any effect? Why or why not?

11 For the Teacher

Objective

In this activity, students must determine why the same bottle leaks in one configuration but does not leak in another.

Materials Needed

- 1 plastic pop bottle
  (The two-liter size works well. You will also need a lid or a solid rubber stopper to seal the bottle.)

- a small nail or compass point to make a hole in the bottle

- a sink, bucket, or dishpan to catch water

Curiosity Hook

You might have a pop container full of colored water sitting where students can see it as they come into the classroom.

Setup

1. Use a small nail or compass point to poke a hole in a plastic pop bottle. Hold your finger over the hole and fill the bottle full of water. Do not put a lid on the bottle. Ask students to predict what will happen when you remove your finger from the hole. Have students write their predictions under 1 on the student page.

2. Now, release your finger. Water should stream out of the hole.

3. Put your finger over the hole in the bottle again. Fill the bottle, and this time, use a lid or a rubber stopper to seal the bottle. Ask students to predict what will happen when you remove your finger from the hole. Have students write down their predictions under 3 on the student page.

4. Now, release your finger. Water should not stream out of the hole (although a few drops may slowly ooze out).

Safety Concerns

Use common sense when poking holes in the bottle.

Outcomes and Explanations

1. The open bottle full of water will leak when you remove your finger from the hole. Why? There are forces working against each other. Air pressure on top of the water and the weight of the water (gravity) on the inside of the hole are pushing against air pressure on the outside of the hole.

Air pressure and gravity pushing out the hole are greater than the air pressure pushing in, so a leak happens. Have students write the explanation under 2 on the student page.

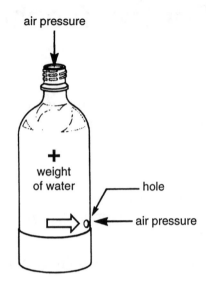

2. The sealed bottle full of water will not leak when you remove your finger from the hole. Why? When you sealed the bottle you eliminated the air pressure on top of the water. Now you have only the weight of the water (gravity) pushing on the inside of the hole and air pressure pushing on the outside of the hole.

The forces on either side of the hole are balanced; thus, nothing moves in or out. Discuss with students how the pressure has equalized, and have them write the explanation under 4 on the student page.

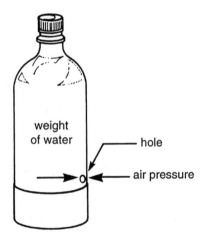

Application

Plastic bottled-water dispensers that pour from a spigot at the bottom of the container utilize the same principle as the above experiment. A hole is punched in the top of the bottle, allowing air to enter. When the spigot is opened, the air pressure above the water plus gravity push the water out the spout.

Take Home

This activity can be easily and safely done by students outside the classroom. However, this experience can get messy, so caution students to use a location where spilling water will not be a problem.

12   Bewildered Berry Basket

Problem: *Is it possible to float a berry basket?*

Predict

1. Predict what will happen when the teacher tries to float Berry Basket A.

Conclude

2. What did happen and why did it happen that way?

Observe

3. Describe what happens when the teacher tries to float Berry Basket B.

Conclude

4. Why did Berry Basket B behave differently from Berry Basket A? List as many reasons as you can think of to explain this difference in behavior.

12   For the Teacher

Objective

In this activity, students will use their powers of prediction and problem solving to determine why one basket floats while the other sinks.

Materials Needed

- 2 plastic berry baskets
  (These can be obtained from the produce section of a grocery store. Only lightweight plastic types will work.)

- 1 clear glass or plastic container large enough to float (and sink) a berry basket
  (A small aquarium works well.)

- a small quantity of liquid dish soap
  (The clear type is preferable.)

- 2 plastic plates
  (White is preferable.)

Curiosity Hook

When students enter the classroom, have the two berry baskets on plastic plates positioned on either side of the float container.

Setup

1. Both berry baskets should be clean and dry.

2. Do nothing to one basket other than place it on a plastic plate. This is Berry Basket A.

3. Rub a small amount of soap on the bottom of the other berry basket and place it on a plastic plate. This is Berry Basket B. Using clear soap and setting the baskets on white plastic plates will help hide the fact that you have altered Berry Basket B.

4. Show students Berry Basket A. Make sure to point out the large holes (openings) in the basket. Ask students to predict what will happen when you attempt to float the basket. Have students write their predictions under 1 on the student page. The basket should float.

5. Now show students Berry Basket B. Attempt to float this basket. This basket should sink. Have students describe the outcome under 3 on the student page.

Safety Concerns

Use common sense.

Outcomes and Explanations

1. How can an object with as many large holes (openings) as a berry basket float? The secret lies in the molecules of the water. Water molecules attract each other and tend to "stick" together (cohesion). When this happens at the surface of the water, the water molecules become more tightly packed and a thin "skin" forms on the surface.

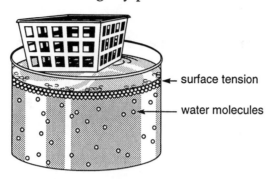

surface tension

water molecules

This thin surface layer ("skin") of tightly packed water molecules is called surface tension. Lightweight objects carefully placed on the surface of calm water will not break this layer and, because of surface tension, will thus float. Discuss surface tension with students and have them write the explanation under 2 on the student page.

2. Why then does Berry Basket B sink? The answer lies with the soap. Soaps and detergents have the property of weakening the cohesive (attracting) forces between water molecules. This weakens the surface tension, allowing the basket to penetrate the surface layer of the water and, therefore, sink. Have students write the explanation under 4 on the student page.

Application

1. Students who have spent time around the edge of a lake or pond may have seen small insects "walking on water" due to surface tension.

2. As demonstrated in this activity, soaps and detergents reduce the surface tension of water. This property makes them useful for washing clothes or your hands. By reducing the surface tension, soaps and detergents make it easier to get the dirt out of your clothes and off your hands.

Take Home

If you place it carefully, even a steel needle can be made to float on water. Challenge students to investigate on their own the effects that weight and shape have on the ability of objects to float due to surface tension.

13 Crazy Card

Problem: *We'll huff and we'll puff, but can we blow the card off the table?*

Predict

1. Predict what will happen to the card when the teacher blows hard beneath it from about 6 inches away.

Conclude

2. Describe what happened. Why did it happen that way?

Predict

3. Now, predict what will happen to the same card when the teacher blows hard beneath it from about two feet away.

Conclude

4. Did changing the distance have an effect on the outcome?

Predict

5. Do you think changing the folding of the card will change the outcome?

Conclude

6. Did changing the shape of the card have an effect?

13 For the Teacher

Objective

In this activity, students will try to determine why the same card stays put one time and blows away the next time (a demonstration of Bernoulli's principle of air and motion).

Materials Needed

- 2 note cards, 5" × 8"
  (Smaller cards will work, but the larger size makes it easier for students to see the results. For visual effect, cut the cards from brightly colored construction paper.)

Curiosity Hook

Have one full-size piece of brightly colored construction paper folded and on display when students come into the classroom.

Setup

1. Fold the cards. Consider the following shapes:

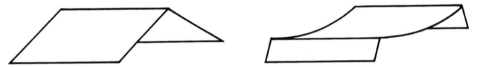

2. Keep both cards out of sight as students come into the classroom.

3. Put one card on a table in view of the students, and ask them to predict what will happen to the card when you blow hard beneath it from about 6 inches away. Have students write their predictions under 1 on the student page. The card should not move and should flatten out.

4. Ask students to predict what will happen when you try the same thing but blow beneath the card from two feet away. Have students write their predictions under 3 on the student page. The card should move and not flatten out. If you place the card near the edge of a table and blow hard enough, the card will skitter off the table.

5. Now, ask students to predict if folding the card in a different shape will change the outcome. Have students write their predictions under 5 on the student page. Use the other card that you folded previously and blow hard beneath it from close-up and then far away. The shape of the card will not alter the effect, so the outcomes should be the same as with the first card.

Safety Concerns

Use common sense.

Outcomes and Explanation

1. When you blow under the card up close, you move the air beneath the card. This lowers the air pressure under the card. The greater air pressure above the card then pushes down and flattens out the card. Discuss this effect with students and have them write the explanation under 2 on the student pages.

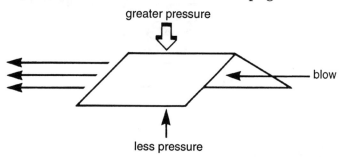

2. When you blow under the card from far away, the air stream loses focus and begins to swirl. This swirling air picks up and moves the card around. Discuss this difference with students and have them write the explanation under 4 on the student page. When you test the second card, the different shape will not affect the outcomes. Have students write this explanation under 6 on the student pages.

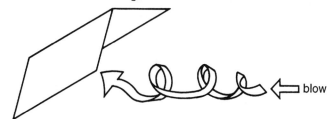

Application

This is an example of Bernoulli's principle which states, The higher the speed of a flowing liquid or gas, the lower the pressure. Give students a 3" × 8" strip of paper. Have them fold over about 1" of one end of the paper. Then have them hold the folded end on their bottom lip and blow gently across the top of the paper strip. The paper strip will rise. Question students as to how we might make practical use of this outcome. We hope that some students will realize that this effect is what allows airplanes to fly. Air moves faster over the curved surface of the top of the wing. According to Bernoulli's principle, as air speed increases, the air pressure on top of the wing drops. The pressure beneath the wing is now greater. A high pressure area always moves towards a low pressure area, in this case pushing upward to create lift beneath the wing.

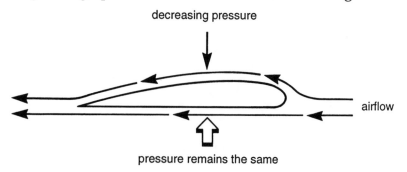

Take Home

Challenge students to experiment on their own outside of class to see if the size or type of material used changes the outcome of this activity.

14   The Diving Peanut

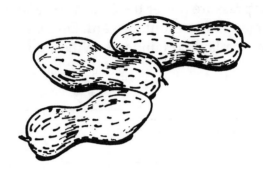

Problem: *Can you sink the peanut?*

Predict

1. In the space below, list as many ways as you can think of to sink the peanut without directly touching it. Sinking involves getting the peanut below the surface of the water at least momentarily.

Conclude

2. What did the teacher do to solve the problem, and how does the teacher's method work?

14 For the Teacher

Objective

In this activity, students will use their powers of critical thinking and problem solving to think of as many ways as possible to sink a peanut without actually touching it.

Materials Needed

- 3 or 4 peanuts in the shell
  (Corks or any small floating objects will also work.)

- a clear plastic or glass container full of water
  (A small aquarium works well.)

Curiosity Hook

Have several peanuts floating in a small aquarium or container where students can see them as they come into the classroom.

Setup

1. Give students several minutes to observe the setup of peanuts floating in a small aquarium or container. You might drop several more peanuts onto the water for effect.

2. Challenge students to think of as many ways as possible to sink the peanuts. The only rule is that they cannot directly touch the peanuts. Have students write and/or diagram their responses to this challenge under 1 on the student page.

3. If no student thinks of this method, I invert a clear plastic or glass cup over one or more of the peanuts and push the cup down into the water. This forces the peanut(s) beneath the level of the water in the larger container without directly touching the peanut(s).

Safety Concerns

Use common sense.

Outcomes and Explanations

1. Discuss this challenge with your class. Accept and list any and all reasonable student solutions. You might actually test some of the student solutions to this challenge.

2. My inverted-cup method works well because air resists being compressed. As you push the inverted cup into the water, the water pushes on the air in the cup, compressing it and increasing the pressure of air in the cup. The air resists and pushes back against the water. This increased pressure prevents water from entering the cup and forces the peanut down beneath the level of the water.

3. Discuss with students the effect of air compression on buoyancy, and have them write the explanation for the demonstration under 2 on the student page.

Application

Many early diving chambers utilized air compression, and it is still used today in places where people are required to work underwater.

Take Home

Encourage students to take this activity from the classroom and use it to teach and amaze their parents, siblings, and friends. Also, challenge students to experiment on their own outside of class to see if changing the size or shape of the container used to sink the peanut affects the outcome of this activity.

15   Off to the Races

Problem: *Which object will win the ramp race?*

Predict

1. Predict which object—a sphere (ball), a disc, or a hoop—will win the race down a short ramp when all the objects are released at the same time.

Conclude

2. Which object was first? second? third?

3. Can you see a pattern to the outcome of the ramp race? Why does it happen that way?

15 For the Teacher

Objective

In this activity, students will use their powers of observation, prediction, and critical thinking to detect patterns in the rolling speeds of different objects.

Materials Needed

- 1 ramp
  (Scraps of paneling or plywood work well. The exact dimensions are not critical, but try for a ramp about 2 feet wide by 4 feet long. Draw a start line on one end of the ramp.)

- assorted spheres: marbles, golf balls, ball bearings
  (The spheres must be solid, not hollow, balls.)

- assorted discs: lids, checkers, flying discs, coasters

- assorted hoops: tape rolls, rings, tires, hula hoops

Curiosity Hook

Have the ramp set up and roll a large object (like a basketball) down the ramp when students enter the room.

Setup

1. Using books or blocks, elevate the ramp about 8 to 10 inches at the end which has the start line on it.

2. Put a piece of tape on the floor about 4 to 5 feet from the lower end of the ramp. Draw a line on the tape to serve as a finish line.

3. Place a sphere, a disc, and a hoop on the start line. At a given signal, let all three objects roll down the ramp towards the finish line.

4. If rolling all the objects at the same time causes problems, try rolling each object separately and timing how long it takes each to cross the finish line. If you use this approach, you will need a stopwatch.

Safety Concerns

Use common sense when rolling objects.

Outcomes and Explanations

The pattern which should become apparent is that all spheres beat all discs, and all discs beat all hoops. The rolling speed is directly related to the distribution of weight around an object's center of gravity. In all three kinds of objects, the center of gravity is the geometric center, but the weights are distributed differently. The closer the weight of an object is to the center of gravity, the faster it can rotate (roll). In the sphere, the weight is most closely distributed around the center of gravity, so it rolls fastest. The hoop has all of its weight located away from the center of gravity, so the hoop rolls the slowest. The disc is somewhere in between. Discuss with students how the weight of an object affects its center of gravity and have them write the explanation under 3 on the student page.

Application

Students may have seen this principle at work when an ice skater goes into a spin. The spin begins with the arms extended. As the spin progresses, the arms are drawn towards the body. This moves the weight closer to the center of gravity, thus causing the skater to spin even faster.

Take Home

Does weight or size make a difference in the outcome of a ramp race? Will a large hoop roll faster than a small disc? Will a heavy disc roll faster than a light sphere? (The answer is no. Spheres always beat discs, and discs always beat hoops regardless of size or weight differences between them. It is the distribution of the weight, not the amount of weight, that is critical.) Challenge students to investigate these questions on their own.

16 Coin in the Cup

Problem: *Can you get the coin into the cup?*

Observe

1. Observe the coin-cup contraption the teacher has set up. In the space below, diagram and describe the coin-cup contraption.

Predict

2. List as many ways as you can think of to get the coin into the cup. There is only one rule—you cannot lift up the card.

Conclude

3. What did the teacher do to solve this problem and how does the teacher's method work?

16 For the Teacher

Objective

In this activity, students will use their powers of observation, critical thinking, and problem solving to offer possible solutions to the problem of getting the coin into the cup.

Materials Needed

- 1 coin
  (A metal washer will also work. I use a penny painted red.)

- 1 clear plastic or glass cup
  (The cup should be heavy enough to remain steady when the card is moved.)

- 1 note card
  (The card must be large enough to cover the opening of the cup.)

Curiosity Hook

Have the coin-cup contraption set up where students can see it as they come into the classroom.

Setup

1. The coin-cup contraption should look like the diagram below:

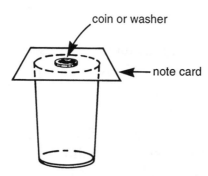

2. Give students several minutes to observe the coin-cup contraption. Have students describe this contraption under 1 on the student page. Then challenge students to think of as many ways as possible to get the coin into the cup. The only rule is that they cannot lift the card. Have students write and/or diagram their responses to this challenge under 2 on the student page.

3. If no student thinks of this method, I flick the card with my finger and the coin falls into the cup. You could also strike the protruding edge of the card with the sharp edge of a kitchen knife. The removal of the card has to be done with a quick action; otherwise, friction will alter the effect you are trying to achieve.

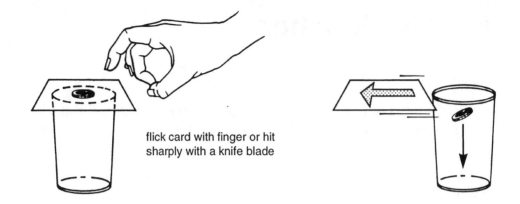

flick card with finger or hit
sharply with a knife blade

Safety Concerns

Use common sense.

Outcomes and Explanations

1. Discuss this challenge with your class. Accept and list any and all reasonable student solutions. You might actually test some of the student solutions to this problem.

2. My flick-the-card method works because objects tend to resist any change in motion. Motionless objects resist moving, and moving objects resist stopping. This property is called *inertia.* The coin lies motionless (inert) on the card. When the card is suddenly flicked away, the coin remains motionless and then drops into the cup.

3. Discuss the property of inertia with students and have them write the explanation under 3 on the student page.

Application

Students have seen inertia in action if they have ever observed a magician jerk a tablecloth off a table full of dishes while the dishes remain in place. Students have felt inertia in action if they have ever been standing on a bus when it suddenly accelerated forward, causing them to stumble backward.

Ask students if it is possible (as movies and television would have them believe) to have people standing motionless and remain standing motionless in a spaceship that suddenly accelerates to fantastic speeds. No, this is not possible, at least not with present technology. In reality, the crew of a spaceship accelerating to light speed would be thrown backward so violently they would not only be killed instantly but most likely atomized. Science fiction writers try to cheat inertia by inventing fictional devices like "inertial dampeners."

Take Home

Encourage students to take this activity from the classroom and use it to teach and amaze their parents, siblings, and friends.

17 Fill It Up

Problem: *Will the jar overflow when the ice melts?*

Predict

1. This jar full of ice will be filled to the rim with hot water. Will the water overflow the jar as the ice melts? Write your prediction in the space below.

Conclude

2. What did happen to the level of the water in the jar when all the ice melted?

3. Why did it happen that way?

17 For the Teacher

Objective

In this activity, students will use their powers of observation, prediction, and critical thinking to understand why a jar full of ice and water will not overflow as the ice melts.

Materials Needed

- 1 jar of clear plastic or glass
- ice
  (Crushed ice works best, as you need to pack the jar as full of ice as possible.)
- hot water
  (Prepare a larger quantity than needed to fill the jar.)
- 1 white paper or plastic plate

Curiosity Hook

Have the jar packed full of ice and sitting in view on a white plate as students come into the classroom.

Setup

1. Pack the jar as full of ice as possible, and set the jar on a white paper or plastic plate. Inform the students that the plate is to catch any water that overflows the jar.

2. Now, show students a container of hot water. Ask them to predict what will happen to the level of the water in the jar when the hot water is added and the ice starts to melt. Have them write their predictions under 1 on the student page. I add some food coloring to the hot water for a more visual effect.

3. Fill the jar of ice to the very rim with hot water, but be careful not to overflow it. For visual effect and precision, I use a dropper to top off the jar of ice with hot water just to the point of overflow.

4. As the ice melts, the resulting increase in liquid water will not overflow the jar.

Safety Concerns

The water does not need to be boiling hot for this activity. Warm water from the tap will suffice and will certainly be safer to work with. Of course, the hotter the water, the faster the effect.

Outcomes and Explanations

Why doesn't the jar overflow as the ice cubes melt? For one thing, the jar contains chunks of ice rather than a single solid piece of ice. This means there are a lot of empty spaces around the chunks of ice. Also, solid water (ice) is strange stuff. It occupies more space and is less dense than liquid water. Therefore, as the ice melts there is more than enough space in the jar to hold the resulting liquid. Discuss this with students and have them write the explanation under 3 on the student page.

Application

Why does ice float? In winter, why does ice form on the top of a pond and not on the bottom? It has to do with density.

$$\text{Density} = \frac{\text{mass}}{\text{volume}}$$

When water freezes it expands. This increases its volume (space occupied), which in turn lowers its density to the point where it is slightly less dense than liquid water. Being less dense than liquid water, ice floats. Why does only a small part of an iceberg stick out of the water? Again, the answer is density. Ice is only slightly less dense than liquid water, so it just barely floats in water with little of the ice sticking out.

Take Home

Encourage students to take this activity from the classroom and use it to teach and amaze their parents, siblings, and friends.

18 Disaster on the Moon

Problem: *Can you survive a crash on the moon?*

You are the captain of an exploratory spaceship that has been surveying the moon. Because of an engine failure, your ship has crashed. You are on the dark side of the moon about 300 kilometers (190 miles) from your mother ship, which is on the light side of the moon. During the crash, some of your crew were injured and much of your equipment was damaged. Your radio failed just before the crash, so no one knows of your problem. Your survival depends on getting to the mother ship as soon as possible.

On the next page is a list of items that were not damaged in the crash. Your task is to rank them in order of importance to your survival. Put a 1 beside the most important item, a 2 beside the next most important, and so on through 15, the least important. Also, list your reasons for the rank you gave each item.

Rank	Item	Reasons
	Box of matches	
	Concentrated food	
	50 feet of rope	
	Solar heater	
	A .45 caliber pistol	
	2 cases dehydrated milk	
	Undamaged space suits for the entire crew	
	2 tanks of oxygen	
	Map of the moon	
	Life raft (self-inflating)	
	Compass	
	5 gallons of water	
	Chemical signal flares	
	First-aid kit	
	Solar radio transmitter	

18 For the Teacher

Objective

In this ranking activity, students will use their powers of critical thinking and their own personal experiences to develop a survival plan for coping with a lunar accident.

Materials Needed

- chart on student page

Curiosity Hook

Display some of the items from the list on a table where students can see them as they enter the classroom. You might want to include photos of the moon or a lunar landing.

Setup

1. Help students imagine themselves in the scenario given at the beginning of the student page.

2. Have students rank the items from 1 (most important) to 15 (least important), and give reasons for their ranking.

3. Students could work on this individually or in small groups.

4. Have students or groups share and defend their ranking. It may cause some consternation among your students, but there are no absolute right or wrong answers. You are looking for soundness of reasoning, not correctness.

Safety Concerns

Students can get argumentative.

Outcomes and Explanations

NASA space survival experts rank the items as listed on the following page.

Application

Analyzing all the variables in a given situation is an important step in the scientific process. Each fact must be weighed and prioritized. Such analysis also exposes students to the idea that there can be more than one "right" answer in a given situation.

Take Home

Students might find it fun and interesting to share this activity with their parents, siblings, and/or friends and see how well they perform as space survival "experts."

Rank	Item	Reasons
1	Space suits	Most vital for crew survival
2	Oxygen tanks	To replace oxygen in space suits
3	Water	To prevent dehydration
4	Map	Needed for navigation
5	Food	To maintain body energy
6	First-aid kit	To treat injured crew members
7	Life raft	For shelter and carrying injured
8	Rope	Tie things for carrying; getting over or across obstacles
9	Flares	To attract attention of mother ship
10	Radio transmitter	Would work once you got to the light side of the moon
11	Solar heater	Would work once you got to the light side of the moon
12	Pistol	Possible self-propulsion device in moon's low gravity
13	Compass	Moon's magnetic field differs from Earth's
14	Milk	Water is too precious to waste mixing it with powdered milk
15	Matches	No oxygen so they wouldn't work

19 Gravity Paper

Problem: *Will the paper fall or fly?*

Predict

1. Predict what will happen when the teacher drops a single sheet of ordinary paper.

Conclude

2. What did happen, and why did it happen that way?

Predict

3. Predict what will happen when the teacher drops a book by itself.

Conclude

4. Is there a difference in the way the book fell?

Predict

5. Predict what will happen when the teacher places the paper on top of the book and drops them both together.

Conclude

6. What did happen, and why did it happen that way?

Predict

7. Will the same thing happen if the paper is larger than the book?

Conclude

8. Did the size of the book in relation to the paper affect the outcome of the experiment? Why?

19 For the Teacher

Objective

In this activity, students will use their powers of prediction and critical thinking to discover why the paper flutters down in one situation, while in another situation it falls like a rock.

Materials Needed

- a sheet of paper
  (The exact size of the paper is not critical.)

- 2 books
  (The size of the books is critical. Book A must be about 1 inch wider and longer than the piece of paper. Book B should be about 1 inch narrower and shorter than the piece of paper.)

Curiosity Hook

As students enter the classroom, the first thing they should see is you about to drop a book from one hand and a sheet of paper from the other hand.

Setup

1. Hold the sheet of paper flat and in front of you at shoulder height. Ask students to predict how the paper will behave when you drop it. Have students write their predictions under 1 on the student page.

2. Now, drop the paper. It will flutter slowly to the floor.

3. Hold Book A flat and in front of you at shoulder height. Ask students to predict how the book will behave when you drop it. Have students write their predictions under 3 on the student page.

4. Drop the book. It will fall straight down rapidly and hit the floor with a bang.

5. Put the paper on top of Book A and hold them flat and in front of you at shoulder height. Ask students to predict how both the paper and the book will behave when you drop them. Have students write their predictions under 5 on the student page.

6. Drop the paper and the book together. As before, the book will fall rapidly straight down. However, the paper will not flutter away but will behave as if it is "stuck" to the book and fall with the book.

7. Put the paper on top of Book B and hold them flat and in front of you at shoulder height. Ask students to predict how both the paper and the book will behave when you drop them. Students should be able to see that the paper sticks out over the edges of the book. Have students write their predictions under 7 on the student page.

8. Now, drop the paper and the book together again. The book will behave as before, but now the paper will peel off the book and slowly flutter down to the floor.

Safety Concerns

Watch your toes.

Outcomes and Explanations

1. When the paper is dropped by itself, it flutters slowly down because the paper is so lightweight and the air the paper is dropping through offers great resistance. Have students write the explanation under 2 on the student pages.

2. When Book A is dropped by itself, it is heavy enough to overcome the resistance of the air and falls in that short distance as if there were no air resistance. Discuss with students the difference in how the book and the paper fell, and have them write the explanation under 4 on the student pages.

3. When the paper is placed on top of Book A and they are dropped together, the paper seems "stuck" to the book, and both plummet rapidly to the floor as a single unit. Why doesn't the paper flutter separately to the floor? It is because there is no air under the paper and, thus, no resistance. It's as if the paper fell in a vacuum. Also, some of the air swirling around the book as it falls may help hold the paper on the book. Discuss this lack of air resistance with students and have them write the explanation under 6 on the student pages.

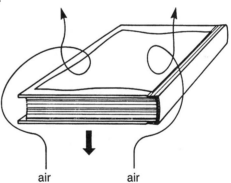

air   air

4. When the paper is placed on top of Book B and they are dropped together, the paper leaves the book because the edges of the paper offer resistance to the air. This air resistance pops the paper off the book, and it slowly flutters down as it did in the first situation. Discuss this effect of air resistance with students and have them write the explanation under 8 on the student pages.

air   air

Application

Have students imagine that they are engineer Farley Flaps. They have been hired to design a new plane that will fly several thousand miles per hour. What shape would they make their plane and why? (The shape should be streamlined to offer as little resistance to the air as possible.) Ask students to offer examples of streamlining. (Cars and airplanes are some examples students should be familiar with.) Students should also understand that it is especially critical for ships and submarines to be streamlined because water is much denser than air and, thus, offers more resistance than air.

Take Home

Encourage students to take this activity from the classroom and use it to teach and amaze their parents, siblings, and friends.

20 Cartesian Diver

Problem: *How does a miniature submersible work?*

Observe

1. Carefully observe and describe the contraption you see before you.

Predict

2. Predict what will happen to the diver when the bottle is squeezed.

Conclude

3. What did happen and why did it happen that way?

20 For the Teacher

Objective

In this activity, students will use their powers of observation and critical thinking to offer possible explanations as to how a Cartesian diver works. Many versions of what has come to be known as the Cartesian diver (named after French scientist Rene Descartes) have been constructed over the years.

Materials Needed

- 1 glass medicine dropper
  (This will serve as the diver. A dropper with a glass tube and a rubber bulb are required. Plastic droppers are so lightweight they will not work. Commercial divers may be purchased from scientific supply companies.)

- 1 clear plastic bottle with lid
  (I recommend a two-liter pop bottle.)

- enough water to fill the plastic bottle

Curiosity Hook

Have the complete diver setup in view as students come into the classroom.

Setup

1. Fill the plastic bottle completely full. Make sure there are no air bubbles trapped inside.

2. Draw some water up into the dropper and place the dropper into the bottle. You want the dropper to just barely float. Add or remove water in the dropper until you achieve the proper buoyancy.

3. Top off the plastic bottle to replace any water lost when adjusting the buoyancy of the diver, and screw on the lid. The complete diver setup should look like this one.

   Have students observe, diagram, and label the complete diver setup under 1 on the student page then write their predictions under 2.

4. Gently squeeze the bottle. The diver will begin to descend. The harder you squeeze, the deeper it goes. When you stop squeezing, the diver ascends. By carefully regulating how hard you squeeze, you can cause the diver to descend and remain at any depth you choose. Under 3 on the student page, have students describe the action of the diver.

Safety Concerns

Use common sense.

Outcomes and Explanations

The pressure you apply to the sides of the bottle is transferred to the water inside. This increased pressure on the water inside the bottle pushes water up into the dropper. As the water level inside the dropper rises, it squeezes the air in the bulb into a smaller space. This increases the weight of the dropper while decreasing its buoyancy, so the dropper sinks.

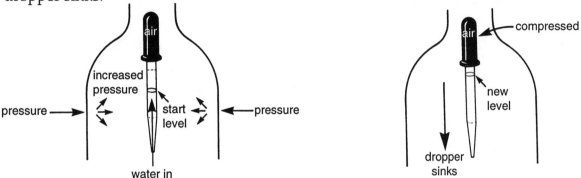

When the pressure on the sides of the bottle is released, the air in the bulb of the dropper squeezes some of the water out of the dropper. The dropper is now lighter and more buoyant and so it rises.

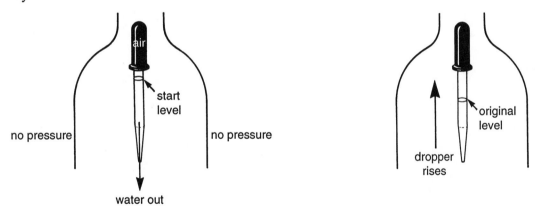

Discuss the effect of air pressure on buoyancy and have students write the explanation under 3 on the student page.

Application

A submarine works under a similar principle. When a submarine dives, special tanks inside the submarine are filled with seawater to increase its weight. The increased weight allows the submarine to sink. In order for the submarine to resurface, compressed air is pumped into the tanks, forcing out the water. This lightens the submarine's weight and increases its buoyancy, allowing it to rise.

Take Home

Students can take this activity from the classroom and use it to teach and amaze their parents, siblings, and friends; the activity works best using the glass dropper with a rubber bulb. However, a variation of the diver can be constructed at home using a glass or plastic jar, a plastic pen top, clay or paper clips, a rubber band, and a balloon. Fill the jar with water. Weight the pen top with the clay or paper clips and place it open-end down so it just floats above the water. Cut the balloon and stretch it over the top of the jar; seal it tightly with the rubber band. Now, press down on the balloon, and the pen top will sink. Release your hand, and the pen top will rise. This method is not as effective as the dropper, but will give you a similar result.

21 Tumbling Tube

Problem: *Will the small tube fall out?*

Observe

1. In the space below, diagram and describe the two-tube contraption you see before you.

Predict

2. What will happen when the teacher turns the two-tube contraption upside down and holds on to only the large tube?

Conclude

3. What did happen and why did it happen that way?

Predict

4. What will happen if we try this again but with a much larger outer tube?

Conclude

5. Does the size of the outer tube affect the outcome?

21 For the Teacher

Objective

In this activity, students will use their powers of observation, prediction, and critical thinking to explain why the small inner tube seems to defy gravity and "fall up."

Materials Needed

- 3 test tubes—1 small and 2 large
  (The small tube should just fit into one of the large tubes. The other large tube should be much larger than the small tube. I recommend the following sizes:

  small tube—13 millimeter outside diameter × 100 mm length

  large tube A—16 mm o.d. × 150 mm length

  large tube B—20 mm o.d. × 150 mm length or 25 mm o.d. × 150 mm length

  Consult your chemistry department for help in securing the proper size tubes for this activity.)

- a dishpan or sink

- enough water to fill both large tubes

- a towel to place in the dishpan or sink to cushion the fall of the small tube

Curiosity Hook

Have the small tube floating upright in large tube A (as per directions below). Set both tubes in a beaker or in a test-tube rack where students can see them as they enter the classroom.

Setup

1. Fill large tube A completely full of water. Push the small tube about halfway into the large tube. The small tube will displace some of the water from the large tube. Eventually the small tube will stop sinking and "float." The upward movement of the small tube can be hard to see without some color contrast. To provide some contrast, you can add colored water to the large tube or, even better, paint the small tube a

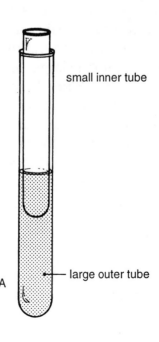

small inner tube

large outer tube

A

bright color. Place the tubes upright in a beaker or in a test-tube rack, and place them where the students can readily observe them. Have students diagram and describe this system under 1 on the student page.

2. Ask students to predict what will happen if you turn this two-tube system upside down and hold on to only the large outer tube. Have students write their predictions under 2 on the student page.

3. Grasp the large tube and turn the whole system upside down, holding on to only the top of the large tube. Water will begin to dribble out of the large tube, and the small tube will slowly move up into the large tube. If the small tube doesn't move, you may have to give it a little upward push.

4. Fill large tube B with water, remove the small tube from large tube A, and place the small tube in large tube B. Ask students to predict what will happen if you turn the two-tube system upside down and hold on to only the large outer tube. Have students write their predictions under 4 on the student page.

5. Place a towel or some other type of cushion in the sink or dishpan because when you invert large tube B, the small tube and the water in the large tube will fall into the sink or dishpan.

Safety Concerns

Be careful handling the glass tubes when wet.

Outcomes and Explanations

1. When large tube A is inverted, the small tube seems to "fall up." This happens because the water dribbling out of the large tube decreases the total volume of water and, at the same time, removes any air above the small tube. Because the tubes fit so tightly, no air can move up into the large tube as the water dribbles out. Since there is no air above the small tube, the outside air pressure pushes the small tube up into the large tube. Explain the outcome of demonstration 1, and have students write the explanation under 3 on the student page.

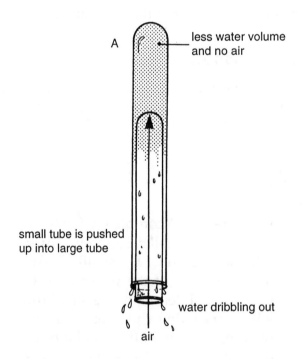

A — less water volume and no air

small tube is pushed up into large tube

water dribbling out

air

2. When large tube B and the small tube are inverted, both the water and the small tube fall out. The size difference between these two tubes allows air to move up into the large tube, equalizing the air pressure on both sides of the small tube, and gravity does the rest. Discuss the change in air pressure with the 2nd tube, and have students write the explanation under 5 on the student page.

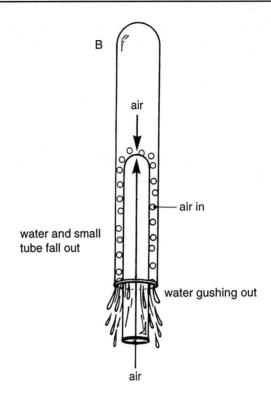

Application

Ask students if air pressure would cause the same effect on the moon. The answer is no. While the moon's gravity might cause water to dribble out of the large tube, there would be no air to push the small tube up into the large tube. Ask students if this same effect would occur in a pressurized space station orbiting the earth. Again the answer is no but for the opposite reason. In this situation, you would have air pressure but relative weightlessness (no apparent gravity) would prevent water from dribbling out of the large tube. And if you don't remove some of the water from the large tube, the small tube cannot be pushed up into the large tube.

Take Home

It may not be practical for students to try this activity outside the classroom. Specially sized tubes are needed for a precise fit, and there is the danger of broken glass from falling or dropped tubes.

22 The Salt and Pepper Mix-up

Problem: *Your little cousin, Daisy Dillrod, has mixed the salt and pepper together.*

Predict

1. Can you think of a way to separate the two substances from each other? Explain your plan below.

Conclude

2. What did the teacher do to solve the problem, and how does the teacher's method work?

22 For the Teacher

Objective

In this activity, students will use their powers of critical thinking and problem solving in an attempt to separate a mixture of salt and pepper.

Materials Needed

- table salt
- finely ground pepper
- a plastic spoon or plastic ruler
- a clear plastic petri dish
- a piece of flannel or wool cloth

Curiosity Hook

Toss some salt over your shoulder as students enter the classroom. Tell students this is to give them luck in trying to solve the challenge you have for them.

Setup

1. Make two mixtures. Put some salt and pepper in the plastic petri dish, and shake the dish to mix them. Then pour some salt and pepper together on a piece of paper and stir to mix them.

2. Challenge students to think of ways to separate these two substances. Have students write their solutions under 1 on the student page.

Safety Concerns

The pepper does fly up, so keep it away from your eyes.

Outcomes and Explanations

1. One solution some students may come up with is to sprinkle the mixture over water. The salt would sink, but the pepper would float on the water where it could be skimmed off. Have students share their solutions to the problem with the rest of the class.

2. My solution to the problem is to employ static electricity.

3. Rub the lid of the petri dish with the flannel or wool cloth, and turn the dish upside down for a short time. Turn the dish right side up and remove the lid. Most of the pepper will be sticking to the lid.

4. Now, rub the plastic spoon or plastic ruler with the flannel or wool cloth, and hold the spoon (or ruler) close to the mixture on the paper. The pepper will literally jump out of the mixture and up onto the charged plastic. (Be careful not to hold the charged plastic too close to the mixture or you will pick up salt as well.)

5. By rubbing the plastic with the flannel or wool cloth, you charge it with static electricity (negative). The pepper has a greater positive charge than the plastic, so it is attracted to the plastic. In electricity and magnetism, like charges repel, but unlike charges attract. Discuss the effects of static electricity with students and have them write the explanation for the outcome under 2 on the student page.

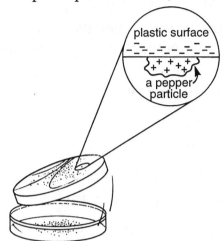

Application

The same kind of electrical attraction often causes dust to gather on your television or computer monitor screen and on charged parts inside.

Take Home

If students have access to the necessary materials, encourage them to take this activity from the classroom and use it to amaze and teach their parents, siblings, and friends.

23 Air: The Heavyweight Champ

Problem: *Does air really have weight?*

Predict

1. Predict what will happen to the stick if the teacher hits the end of the stick.

Conclude

2. What did happen, and why did it happen that way?

Predict

3. The teacher will spread several sheets of newspaper over the stick. Predict what will happen to the stick now if the teacher hits the end of the stick.

Conclude

4. What effect did adding the newspaper have? Why?

23 For the Teacher

Objective

In this activity, students will use their powers of observation, prediction, and critical thinking to understand why the stick breaks when the air "sits" on it.

Materials Needed

- several sticks
  (The dimensions for each stick should be approximately 1/4 inch thick by 1 inch wide by about 2 feet long. Old yardsticks will work, or you might check local lumberyards or hardware stores.)
- a meterstick or cane
- several sheets of newspaper
- safety goggles or glasses

Curiosity Hook

Twirl the meterstick or cane and/or make chopping motions with them as students enter the classroom.

Setup

1. Place a stick on a table with a smooth surface, and let about 2 inches of the stick protrude over the edge of the table.

2. Ask students to predict what will happen to the stick when you strike the protruding end with a meterstick or cane. Have students write their predictions under 1 on the student page.

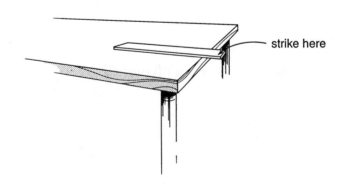

strike here

3. Now strike the protruding edge of the stick with a sudden, sharp blow. The stick will fly off the table.

4. Place the stick back on the table as before and cover it with several double sheets of newspaper.

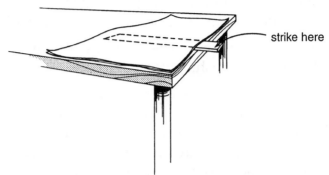

strike here

5. Ask students to predict what will happen to the stick when you strike the protruding end with a meterstick or cane. Have students write their predictions under 3.

6. Again, strike the protruding end of the stick with a sudden, sharp blow. This time the protruding end of the stick will break off and fly through the air.

7. You can repeat this effect several more times by pulling the stick out another 2 inches or so and striking it again.

Safety Concerns

Pieces of broken stick will be flying around during part of this activity so you should do the following:

1. Wear safety goggles.

2. Move students back a safe distance.

3. Orient the stick so the broken pieces will fly away from where the students are located.

To reduce the risk of injury or breakage, you might consider doing this activity in a gymnasium or outside.

Outcomes and Explanations

When you strike the end of the resting stick, the energy (or part of it) from the striking meterstick or cane is transferred to the resting stick, causing it to fly off the table. Discuss this transference of energy with students and have them write the explanation for demonstration 1 under 2 on the student page. When you cover the resting stick with the newspaper, you have a whole column of air pushing down with great weight ("sitting") on the paper and the resting stick. The weight of the air on the newspaper is so great that when you hit the resting stick with the striking stick, the resting stick cannot budge the newspaper and consequently breaks. Discuss the effect the newspaper has on the resting stick, and have students write the explanation for demonstration 2 under 4 on the student page.

Application

Following these steps, have students calculate how much air pressure (weight) is actually pushing down ("sitting") on the newspaper and resting stick.

Step 1—Determine the size of the newspaper by measuring its height and width. (Most newspapers are around 27 inches wide by 23 inches high)

Step 2—Multiply the width times the height to determine the surface area of the newspaper that the air is pushing down on. ($27 \times 23 = 621$ square inches)

Step 3—Multiply the surface area of the newspaper times the weight of the air. Atmospheric pressure at sea level is 14.7 pounds per square inch, so $621 \times 14.7 =$ 9,128.7 pounds. This is close to the weight of two station wagons. The striking stick cannot impart enough force to the resting stick to lift this much weight, so the resting stick has no choice but to break.

Take Home

Students could certainly challenge their parents, siblings, and/or friends with this activity. However, caution them to follow good safety procedures:

1. Do the activity outside. This greatly reduces the risk of injury or breakage when pieces of stick start flying around.

2. The striker should wear safety goggles or glasses.

3. Spectators should remain back a safe distance.

24 Fire in the Hole!

Problem: *Watch out for the exploding can!*

Observe

1. Observe the exploding can contraption the teacher has prepared. In the space below, diagram and describe this contraption.

Conclude

2. Describe what happened when the teacher blew into the rubber tube.

3. Why did it happen that way?

24 For the Teacher

Objective

In this activity, students will use their powers of observation and critical thinking and their past experiences to understand why the can blows its top.

Materials Needed

- a metal can or bucket with a tight-fitting friction lid

- a small funnel

- about 3 feet of rubber tubing

- a small candle

- about one teaspoon of flash powder
  (Suitable flash powders are powdered coffee creamer or lycopodium powder. Lycopodium powder is actually the pollen of a plant and, while it has excellent flash qualities, it must be purchased from a scientific supply company.)

- several books to hold the can

- safety goggles

Curiosity Hook

Following the diagram, have the can set up where students can see it when they come into the classroom. To really grab your students' attention and peak their curiosity, paint the can beforehand with some bright color.

Setup

1. Punch a hole in the bottom of the can to fit the stem of the funnel. Place the funnel into this hole.

2. Set the can on two stacks of books to allow the funnel stem to protrude below the bottom of the can.

3. Attach the rubber tubing to the funnel stem.

4. Pour about 1/4 teaspoon of flash powder into the funnel. The one teaspoon of flash powder called for in the materials section should allow you to repeat this activity about four times. Students will request that you use larger amounts of flash powder, hoping for progressively larger explosions. Larger explosions will happen using up to about 1/2 teaspoon of flash powder. Beyond that amount, diminishing returns occur, and powder is wasted because not all of it ignites.

5. Light a candle and stand it in the can opposite the funnel.

6. Set the lid lightly on the can, leaving a small opening to feed air to the candle. Now, have students diagram and describe this setup under 1 on the student page.

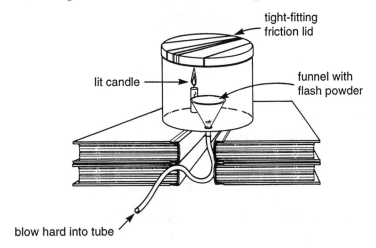

tight-fitting
friction lid

lit candle

funnel with
flash powder

blow hard into tube

7. Relight the candle, if necessary, and close the lid tightly on the can. Have a student quickly turn off the classroom lights to heighten the effect.

8. Blow a sharp, hard puff of air through the rubber tubing. The lid should blow off the can with a whoosh and a flash of orange flame.

Safety Concerns

You will be generating a small flash of flame, so wear safety goggles when blowing into the rubber tubing, and position students a safe distance from the can.

Outcomes and Explanations

When you blow hard into the rubber tubing, you shoot the fine particles of flash powder into the air inside the can. These powder particles are flammable and form an explosive mixture with the oxygen in the can. The candle flame ignites the mixture almost instantly. This explosive combustion produces large amounts of gaseous carbon dioxide (CO_2) and water vapor. The sudden production of these gases pressurizes the inside of the can, blowing off the lid. Have students write the explanation under 3 on the student page.

Application

Depending on the part of the country you live in, students have probably heard of errant sparks causing dust explosions in coal mines or grain bins. The effect is similar to this experiment but on a much larger scale.

Take Home

Lack of proper equipment precludes students doing this activity outside the classroom. Furthermore, there is the risk of flame and fire to consider if students try this activity at home.

25 Slippery Cube

Problem: *Can you pick up an ice cube with a piece of string?*

Predict

1. In the space below, list as many ways as you can think of to pick up the ice cube using only a piece of string. The only rule is you cannot tie the string around the ice cube.

Conclude

2. What did the teacher do to solve this problem, and how does the teacher's method work?

25 For the Teacher

Objective

In this activity, students will use their powers of critical thinking and problem solving to think of as many ways as possible to pick up an ice cube with a piece of string.

Materials Needed

- 1 ice cube

- a jar, glass, or beaker to serve as a stand for the ice cube

- a piece of string or sewing thread about 10 inches long

Curiosity Hook

Have the ice cube on the stand where students can see it as they enter the classroom.

Setup

1. Invert the jar, glass, or beaker to serve as a stand for the ice cube.

2. Set the ice cube on the stand.

3. Hold the string near the ice cube.

4. Challenge students to think of as many ways as possible to pick up the cube using only the string but without tying or wrapping the string around the cube. Have students write and/or diagram their solutions to the problem on the student page under 1.

5. If no student thinks of this method, I soak about 2 inches of one end of the string by placing it in my mouth. I then lay this part of the string on the ice cube. In a few seconds, the string will be frozen to the cube, and I can easily lift the cube off the stand with the string. (NOTE: If the ice cube has begun to melt, it may be more difficult to get the string to adhere. Start with a fresh ice cube from the freezer or shake off any excess water from the display cube before you place the string on top.)

Safety Concerns

Wet floors can be slippery; wipe up any spills.

Outcomes and Explanations

1. Discuss this challenge with your class. Accept and list any and all reasonable student solutions. You might actually test some of the student solutions to this problem.

2. My wet-string method works because the cold temperature of the ice cube freezes the water on the string. As this happens, the string becomes frozen to the ice cube. The string is now physically attached to the cube, and you can easily pick up the cube by pulling up on the string. Have students write the explanation under 2 on the student page.

Application

It is very tempting during the winter for small children to touch their tongues to ice-cold metal. If your students haven't tried this, they probably know someone who has. What will happen to a child who does this and why? (The moisture on the tongue will be changed to ice, freezing the tongue to the metal just as surely as the string in this activity was frozen to the ice cube.)

Take Home

Encourage students to take this activity from the classroom and use it to amaze and teach their parents, siblings, and friends.

26 The Mystery of the Starving Sloths

Problem: *Can you save the starving sloths?*

You are the manager of a large, modern zoo. You would like to exhibit a sloth, a strange plant-eating mammal from the rain forests of Central and South America. You learn that there are two varieties of sloths—two-toed and three-toed—and that both types eat only leaves for food.

You send your collectors into the jungles, and they bring back some three-toed sloths. You put the sloths into the habitat you have prepared, and you feed them leaves from trees around the zoo. However, the three-toed sloths begin to die of starvation, even though an autopsy shows they died with a stomach full of leaves. Why are the three-toed sloths starving? Recent research on sloths may provide some useful clues.

It was once believed that sloths spend their entire lives in one tree. We now know this is not true. Each sloth has a different feeding pattern. A female sloth feeds on over 40 tree species, and she teaches this pattern to her young. The female also passes on special microorganisms to her young by licking them. These microorganisms settle in the young sloth's stomach and intestines, where they help the sloth digest certain types of leaves.

Predict

1. Why are the sloths starving?

Conclude

2. What would you do to keep the sloths well-fed and healthy?

26 For the Teacher

Objective

In this activity, students will use their powers of critical thinking and problem solving to determine why the sloths are starving even though they are being fed.

Materials Needed

None

Curiosity Hook

Have a picture of a sloth where students can see it when they come into the classroom.

Setup

Students should read the scenario and background information on the student page.

Safety Concerns

None

Outcomes and Explanations

1. Discuss the scenario and background information with students.

2. Have students write their explanations for why the sloths are starving and their solutions to the problem on the student page.

3. Why are the sloths starving? The sloths are being fed leaves they do not normally eat, and, apparently, they cannot digest these leaves.

4. How can we keep the sloths well fed? Before the sloths are captured, much time should be spent observing them to record their exact feeding patterns. In captivity, the sloths would have to be provided with the same leaves they were used to eating in the wild. Also, very young sloths should not be captured without also capturing their mother. It is critical that the mother teaches the young what to eat and passes on the needed microorganisms.

Application

Many students have been to a zoo, so this activity could serve as a springboard for a discussion about zoos—Why do we have zoos? Should we have zoos? Are zoos designed for the health and well-being of the animals or the convenience of the people? How could zoos improve? and so on.

27 Beat the Bottle

Problem: *Can your lungs beat the bottle?*

Predict

1. Predict what will happen when someone tries to blow up the balloon.

Conclude

2. Did the balloon inflate? Why or why not?

Predict

3. Predict what will happen when someone tries to blow a wad of paper into a bottle.

Conclude

4. What did happen, and why did it happen that way?

27 For the Teacher

Objective

In this activity, students will use their powers of observation and knowledge of air pressure to determine why, in a balloon-blowing contest, the bottle always wins.

Materials Needed

- a clear plastic two-liter pop container
- a large balloon
- a piece of tissue paper

Curiosity Hook

Have the pop container with the balloon inside set up where students can see it as they come into the classroom.

Setup

1. Put a large balloon down into an empty pop container, and stretch the neck of the balloon completely over the mouth of the bottle. Ask students to predict what will happen when someone tries to blow up the balloon. Have students write their predictions under 1 on the student page.

2. Have a student try to blow up the balloon inside the bottle. Student will not be able to inflate the balloon.

3. Remove the balloon from the pop container, and lay the container on its side or hold it in a horizontal position. Tear a small piece of paper from a tissue and make a wad. Place the wad of paper just inside the opening at the top of the bottle. Ask students to predict what will happen when someone tries to blow the wad of paper into the bottle. Have students write their predictions under 3 on the student page.

4. Have a student try to blow the wad of paper into the bottle. The student should position himself so he is blowing directly into the bottle. The wad of paper will pop out of the bottle and fly towards the person blowing into the bottle.

Safety Concerns

Use common sense.

Outcomes and Explanations

1. To inflate the balloon, you would need to compress the air trapped between the balloon and the bottom of the bottle. Human lungs are not strong enough to inflate the balloon and to compress the trapped air.

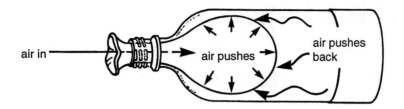

Have students write the explanation under 2 on the student page.

2. Blowing into the bottle increases the air pressure in the bottle. Also, moving air has less pushing power than still air. Since the bottle is open, air pushes out of the bottle, taking the wad of paper with it.

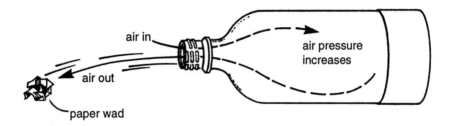

Have students write the explanation under 4 on the student page.

Application

Compressed air has great strength. The compressed air inside a tire can support the weight of a bicycle or a car.

Take Home

Encourage students to take this activity from the classroom and use it to amaze and teach their parents, siblings, and friends.

28 "Paint" the Paper

Problem: *Can you meet the challenge and spray the water?*

Predict

1. In the space below, list as many ways as you can think of to spray the colored water onto a white sheet of paper. There are two rules:

   a. You can only use one straw.

   b. You may bend, cut, or change the straw in any way you choose.

Conclude

2. What did the teacher do to meet the challenge? How does the teacher's method work?

28 For the Teacher

Objective

In this activity, students will use their powers of observation, critical thinking, and problem solving to try to "paint" the paper.

Materials Needed

- several plastic drinking straws
  (Transparent straws are preferred as they will allow students to see water movement in the straw.)
- a container of water colored with food coloring
- several sheets of white paper

Curiosity Hook

Have a container of colored water with a straw in it sitting where students can see it as they come into the classroom.

Setup

1. Have students observe the container of colored water with the straw in it.

2. Challenge students to use the straw to "paint" (spray) colored water onto a piece of white paper. They are limited to one straw, but they may alter the straw in any way they choose. Have students write and/or diagram their responses to this challenge under 1 on the student page.

3. If no student thinks of this method, I make a spray "painter." Cut the straw so one piece is about 3 inches long, and place this piece in the colored water. Hold the second piece at a right angle to the first so that the ends of the two pieces are close together.

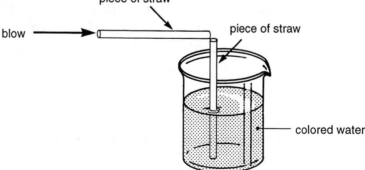

blow → piece of straw   piece of straw   colored water

Have a student hold up a piece of white paper. When you blow through the horizontal tube, colored water should spray onto the paper. The positioning of the ends of the two straws is critical to achieving any results, so practice this ahead of time. You might want to try cutting the straw only partially instead of into two pieces. This might make it easier to align the two parts of the straw. Experiment to see which works best for you.

Safety Concerns

Use common sense when spraying the "paint."

Outcomes and Explanations

1. The easiest way to meet the challenge (and the way most students will think of) is to suck colored water up into the straw and then spray it onto the paper.

2. The spray "painter" described works on Bernoulli's principle of pressure and velocity. As air is blown across the top of the vertical straw, the pressure is lowered. Therefore, the pressure at A is lower than at B. The greater air pressure below pushes the colored water up into the vertical straw. Once the water reaches the top of the straw, the horizontal flow of air sprays it onto the paper.

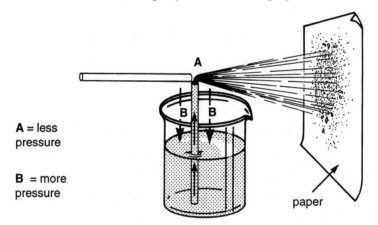

A = less pressure

B = more pressure

paper

3. Discuss with students how Bernoulli's principle applies to this activity, and have them write the explanation under 2 on the student page.

4. Extend this activity by having students predict what the effect would be of lengthening the vertical straw. (The longer the vertical straw, the harder you have to blow to achieve the effect. If you keep increasing the length of the vertical straw, you reach a point at which the pressure difference is not great enough to force water to the top of the vertical straw.)

Application

Applications of this principle are found in spray-painting guns, spray bottles, and pressurized spray cans.

Take Home

Encourage students to take this activity from the classroom and use it to amaze and teach their parents, siblings, and friends.

29 The Mystery of the Sunken Cube

Problem: *Can you solve the mystery of the sunken cube?*

Observe

1. Carefully observe and accurately describe the situation you see before you.

Conclude

2. What is going on here? Since the containers seem identical, what variables (differences) could be affecting the outcome? In the space below, list as many reasons as possible to explain the mystery of the sunken cube.

3. Why did one cube sink and one float? What is the answer to the mystery of the sunken cube?

29 For the Teacher

Objective

In this activity, students will use their powers of observation and critical thinking to solve the discrepancy between the sunken cube and the floating cube.

Materials Needed

- 2 ice cubes

- 2 glass or plastic containers
  (Glasses, cups, beakers, or jars will suffice, but they must be transparent in order for students to see the ice cubes. The containers should be as similar as possible to eliminate the size, shape, or composition of the containers as variables.)

- 2 covers for the containers
  (The containers need not be sealed tightly but only have some covering over the top to prevent students from smelling the alcohol in container B. Jar lids, note cards, or squares of glass or plastic will work, depending on the type of container used.)

- enough water to fill container A

- enough alcohol to fill container B
  (Use rubbing [isopropyl] alcohol. It works well and is relatively cheap and easy to obtain.)

- food coloring

Curiosity Hook

Have the containers sitting where students can see them as they come into the room. Once the students get settled in, drop an ice cube into each container.

Setup

1. Freeze two ice cubes ahead of time. To add another variable to this activity, use two different-colored ice cubes.

2. Fill one container nearly full of water, and label this container A.

3. Fill the other container nearly full of rubbing alcohol, and label this container B. Both containers should be filled with identical amounts of liquid to eliminate the amount of liquid as a variable.

4. When the cubes are added, the ice cube in water will float, but the cube in alcohol will sink. Keep the lids on the containers so students cannot smell the alcohol in container B.

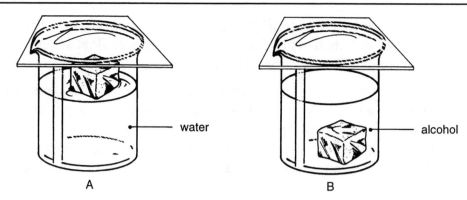

water

alcohol

A   B

5. Have students carefully observe the two containers and then accurately describe the situation they see. Have them diagram and write their descriptions under 1 on the student page.

6. Now, have students come up with as many hypotheses as possible to explain the mystery of the sunken cube. Have them write their hypotheses under 2 on the student page. Students should understand that what they are listing are variables (differences) between the two containers.

Safety Concerns

Rubbing alcohol can cause serious gastric disturbances if taken internally. Dispose of alcohol by pouring it on a sidewalk and letting it evaporate.

Outcomes and Explanations

1. Discuss student hypotheses (explanations) as to why one cube sinks and one cube floats.

2. If the containers are identical and the amount of liquid in both containers is nearly identical, the only variables (differences) between the two containers are as follows:
   a. the liquids are different
      (Just because both liquids are clear doesn't mean that both liquids are water.)
   b. the color of the ice cubes
   c. the composition of the cubes
      (If students take for granted that the cubes are composed of only water, suggest that this might not be the case.)
   d. the temperature of the liquids

3. The answer to this discrepancy lies in the liquids. Alcohol is less dense (compact) than ice, so the ice sinks in alcohol. Water is more dense (compact) than ice, so the ice floats in water. Discuss this discrepancy with students and have them write the explanation under 3 on the student page.

Application

As seen in this experiment, the density of a material is one factor in its ability to float or sink in a liquid. Most substances will contract in volume when frozen, so the density of the solid is higher than the density of the liquid. Water behaves differently. It expands when it freezes, and its density decreases. Thus, the ice floated on the water because substances with a lower density float in liquids with a higher density.

Density is defined as the mass per unit volume of a material. Density tells us how much mass there is in a certain volume of something. Density is expressed as grams per cubic centimeter (g/cm³); for example, the density of lead is 11.3 g/cm³, while aluminum has a density of 2.7 g/cm³. Thus, a piece of lead has a much greater mass than a piece of aluminum with the same volume. Lead feels heavier and is much more compact than aluminum.

Take Home

There are safety factors to consider if students attempt to conduct this activity outside the classroom. Rubbing alcohol can cause gastric problems if ingested and is highly flammable. Furthermore, there is no way to ensure that students will dispose of the alcohol according to safety and environmental guidelines.

30 Transfer the Water

Problem: *Can you meet the challenge and transfer the water?*

Predict

1. In the space below, list as many ways as you can think of to transfer the water from the full container to the empty container. There are two rules:

    a. You cannot move either container over or around the barrier.

    b. You may lift either container up into the air, but it must remain on its own side of the barrier.

Conclude

2. What did the teacher do to solve the problem? How does the teacher's method work?

30 For the Teacher

Objective

In this activity, students will use their powers of observation, critical thinking, and problem solving to meet the challenge of transferring the water.

Materials Needed

- 2 containers
  (Shape is not critical, but they should be small enough for you to easily pick up and pour with.)

- a barrier of some type to place between the containers

- enough water to fill one container 3/4 full
  (Adding food coloring to the water will make the results more visible to students and generate even more curiosity.)

- 1 piece of string about one foot long
  (The string must be able to absorb water. Do not use nylon fish line.)

Curiosity Hook

Have one container of water and one empty container set up with the barrier between them so students can see this as they enter the classroom. You may display the string or keep it out of sight, depending on how much direction you want to give the students.

Setup

1. The initial setup should look like this:

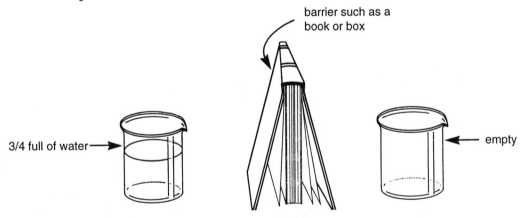

2. Give the students several minutes to observe the setup, and then challenge them to think of as many ways as possible to transfer the water from the full container to the empty container. The rules allow them to lift the container(s) on each side of the barrier, but the rules do not allow them to move the containers over or around the barrier. Have students write and/or diagram their responses to this challenge under 1 on the student page.

3. If no student thinks of this method, I go ahead and demonstrate it. (If a student does come up with this method, have him/her help you with the demonstration.) First, I wet the string completely and then slowly pour the water down the string. It works best if the string is held fairly taut.

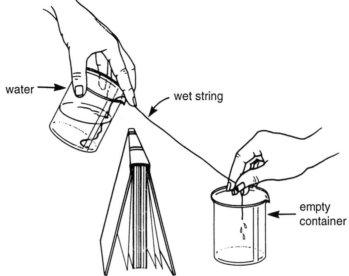

water →   wet string

empty container

Safety Concerns

Use common sense.

Outcomes and Explanations

1. Discuss this challenge with your class. Accept and list any and all reasonable student solutions. You might actually test some of the student solutions to this challenge.

2. Pouring water down the string works because the water molecules "stick" to the molecules of the string (adhesion). Once the string is wet, the water clings to the water molecules already present (cohesion) and follows them down the string.

3. Discuss the effects of adhesion and cohesion with students and have them write the explanation under 2 on the student page.

Application

Other materials that have the same water-absorbing properties as string, like cotton, cloth, paper, or wood, would also work. Materials like nylon and wool, which are not water-absorbent, would not work. Have students think of ways to apply the principles of adhesion and cohesion to activities in their daily lives. One example would be using a paper towel to absorb spilled liquid (adhesion).

Take Home

Encourage students to take this activity from the classroom and use it to teach and amaze their parents, siblings, and friends. Also, challenge students to experiment on their own outside of class to see if the type of liquid (vinegar, syrup, etc.) changes the outcome of this activity.

31 | The Floating Paper Clip

Problem: *Can you solve the mystery of the floating paper clip?*

Observe

1. Carefully observe the floating paper clip contraption and, in the space below, accurately describe the situation you see before you.

Conclude

2. List as many hypotheses (explanations) as possible to explain the mystery of the floating paper clip.

3. What is the answer to the mystery of the floating paper clip?

31 For the Teacher

Objective

In this activity, students will use their powers of observation, critical thinking, and problem solving to solve the mystery of the floating paper clip.

Materials Needed

- 1 magnet
  (The more powerful the magnet, the better. I borrow one from the physical science/ physics department.)

- 1 clamp

- 1 rod and stand

- 1 paper clip

- a short length of thread or light string

- several pieces of cardboard or a shoe box to make a shield to hide the magnet

Curiosity Hook

Have the mystical, magical floating paper clip contraption set up where students can see it as they enter the classroom.

Setup

1. The setup should look like this:

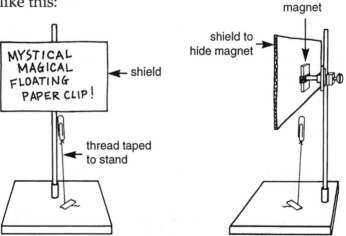

Students should not be able to see the magnet. The stronger the magnet, the greater the gap you can have between the clip and the magnet and, thus, the better the illusion. An alternative to this setup is to anchor a pencil in some clay. Obtain two or more ring magnets with holes in the center. When the ring magnets, with opposite poles facing each other, are placed on the pencil, they will levitate.

2. Have students observe the setup and then accurately describe the situation they see. Have them write and diagram their descriptions under 1 on the student page.

3. Now, have students come up with as many hypotheses as possible to explain the mystery of the floating paper clip. Have them write their hypotheses under 2 on the student page.

Safety Concerns

Use common sense.

Outcomes and Explanations

1. List and discuss the various student hypotheses offered to explain the floating paper clip.

2. To show students that nothing is holding up the clip, slide a thin, nonmetallic object such as a note card, plastic comb, plastic ruler, etc. through the gap between the magnet and the clip.

3. Remove the shield around the magnet, and show students that the answer to this mystery lies with the invisible lines of force we call magnetism. Now, take a pair of metal scissors and "cut" the invisible magnetic lines between the clip and the magnet. (As the scissors close, they will break the magnetic attraction.)

4. Magnetism is generated by spinning electrons within atoms. Each moving electron forms a magnetic field around itself. In most atoms, each electron is paired with another electron spinning in the opposite direction. Thus, the magnetic field of one electron is cancelled by the magnetic field of the other electron. In some materials, such as iron, cobalt, and nickel, the atoms have unpaired electrons, so the magnetic field is not cancelled. As a result, each atom acts like a small magnet, causing the entire material to generate invisible lines of force we call magnetism. Certain materials, such as the paper clip, are attracted by these lines of force. Discuss the principle of magnetism with students and have them write the explanation for the floating paper clip under 3 on the student page.

Application

1. Have students list and discuss ways magnetism makes an impact on their lives. Magnetism is commonly used in electric motors and compasses and in the generation of electricity.

2. Students can make a magnetic-field picture. Using a hot plate, have them melt candle wax or paraffin in a large metal pan. They should dip sheets of paper into the liquid and then hang the papers up and allow them to cool. The coated, cooled paper should be placed on a piece of cardboard. Next, students should add iron filings and design a picture on the coated paper using a magnet. Lastly, the cardboard and coated paper should be placed on a warm (not hot!) hot plate until the wax on the paper softens. The iron filings will sink into the soft wax, forming a permanent picture.

32 Heat Racers

Problem: *Which student will win the heat race?*

Observe

1. Observe the two heat racers the teacher is holding. In the space below, accurately describe the heat racers.

Predict

2. Two students will now hold the heat racers in a flame. The wire that gets too hot to handle will be the winner. Predict which student will win the heat race.

Conclude

3. What is going on? Why does one wire get hot but not the other? List as many hypotheses as possible to explain this.

4. What is the reason one wire gets hot but not the other?

32 For the Teacher

Objective

In this activity, students will use their powers of observation and critical thinking to apply the principle of heat conduction to the mystery of why one wire gets too hot to hold while the other wire remains cool.

Materials Needed

- 3 pieces of heavy copper wire
- 2 large corks
- a heat source, such as a Bunsen burner or an alcohol burner
- 2 pairs of safety goggles

Curiosity Hook

Hold the heat racers and have the burner going when students enter the classroom.

Setup

1. The heat racers should be constructed as follows:

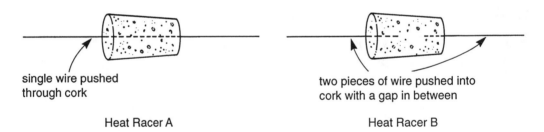

single wire pushed
through cork

two pieces of wire pushed into
cork with a gap in between

Heat Racer A   Heat Racer B

2. Have students carefully observe the two heat racers and then accurately describe them under 1 on the student page.

3. Ask two students to hold the heat racers in the flame. Since an open flame is involved, students should wear safety goggles.

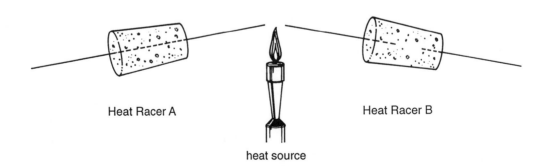

Heat Racer A   Heat Racer B

heat source

4. Ask students to predict which student will win the heat race. That is, which wire will become hot first? Have students write their predictions under 2 on the student page.

5. In a short time, Heat Racer A gets too hot to hold, but Heat Racer B remains cool to the touch.

6. Have students come up with as many reasons as possible to explain the discrepancy in heating between the two racers then write their reasons under 3 on the student page.

Safety Concerns

The students holding the wires should wear safety goggles. Remind them to let go of the wire as soon as it feels hot. Provide a safe place such as an aluminum pie pan to place the wire that gets hot.

If you are concerned about the safety of the students holding the wires, either you or the students can hold the racers by the cork only. Put one end of the wire in the flame and then touch the other end to a thermometer. It only takes a minute or two for the wire to heat up. Be sure to touch Heat Racer A to the thermometer just long enough to indicate a temperature change; there will be no temperature change in the wire of Heat Racer B. If you hold the hot wire against the thermometer too long, the glass bulb might break from the heat.

Outcomes and Explanations

1. Discuss this discrepancy in heating with your class. Accept and list any and all reasonable student solutions.

2. Diagram the internal structure of both heat racers. Heat Racer A gets hot because the heat from the flame is conducted to the end where the wire was held; however, the heat in Heat Racer B is conducted only to the cork. The cork absorbs the heat and insulates the end of the wire being held. Thus, it remains cool. Have students write this explanation under 4 on the student page.

Application

Have students discuss and list examples of heat conduction and insulation in everyday life. One example would be using a hot pad to handle hot pans in the kitchen.

Take Home

Because of the fire hazard of an open flame and the possibility of being burned by a hot wire, it is best not to have students try this activity outside the classroom.

33 Go with the Flow

Problem: *Which way will the water flow?*

Predict

1. In the space below, predict what will happen in each set of bottles when the card between them is removed.

Conclude

2. Was there any difference in what happened with the first set of bottles and the second set? If so, describe what happened that was different. What would cause this?

33 For the Teacher

Objective

In this activity, students will use their powers of observation and critical thinking to understand why water flows in one situation but not the other.

Materials Needed

- 4 empty glass or heavyweight-plastic containers of identical size (Some of the newer soda bottles are too lightweight to handle easily. Remove the labels so that students can better observe the water flow.)
- 2 note cards
- food coloring
- both warm and cold water

Curiosity Hook

Have the bottles and cards set up where students can see them as they enter the classroom.

Setup

1. Set up the bottles as follows:

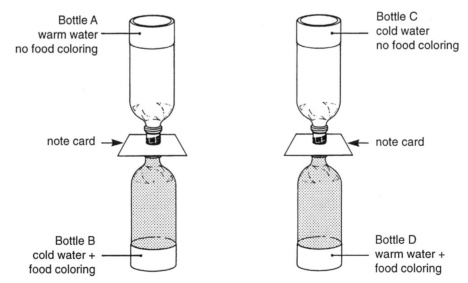

Bottle A
warm water
no food coloring

note card →

Bottle B
cold water +
food coloring

Bottle C
cold water
no food coloring

← note card

Bottle D
warm water +
food coloring

2. Have students carefully observe the bottles. Ask students to predict what will happen in each set of bottles once the cards are removed. Have students write their predictions under 1 on the student page.

3. Carefully remove the card between bottles A and B so that the bottles remain aligned and open to each other. There should be no water movement between bottles A and B.

4. Now, carefully remove the card between bottles C and D so that the bottles remain aligned and open to each other. The colored water in bottle D should flow up into bottle C.

Safety Concerns

Wet floors are slippery; clean up any spills.

Outcomes and Explanations

1. Discuss this discrepancy in water flow with your class. Accept and list any and all reasonable student solutions.

2. The answer to the discrepancy in water flow between the two sets of bottles lies with the temperature difference between the bottles. Cool or cold water and air sink because they are denser; warm or hot water and air rise because they are less dense. In bottles A and B, the cold water is already on the bottom and the warm water is on the top, so no movement occurs. In bottles C and D, however, the denser cold water is on top, so it sinks, and the less dense warm water is on the bottom, so it rises. Discuss with students the effect of temperature on water flow, and have them write the explanation under B on the student page.

Application

Ask students where in nature such movements due to temperature differences occur. Examples of such movement would be the wind and ocean currents.

Take Home

Encourage students to take this activity from the classroom and use it to teach and amaze their parents, siblings, and friends.

34 Hang On!

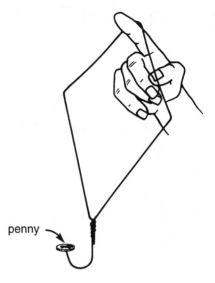

penny

Problem: *Will the penny hang on or fly off?*

Predict

1. In the space below, predict what will happen to the penny when the teacher begins to swing the contraption hanging from his/her finger.

Conclude

2. What happened to the penny? Why did it happen that way?

34   For the Teacher

Objective

In this activity, students will use their powers of observation, prediction, critical thinking, and problem solving to understand why the penny "sticks" to the coat hanger.

Materials Needed

- 1 wire coat hanger
  (NOTE: You may need to file the end of the curved portion to get the penny to lie flat on the wire.)

- 1 penny

- safety goggles

Curiosity Hook

Swing the wire coat hanger contraption (without penny) on your finger as students enter the classroom.

Setup

1. Bend the wire coat hanger as follows:

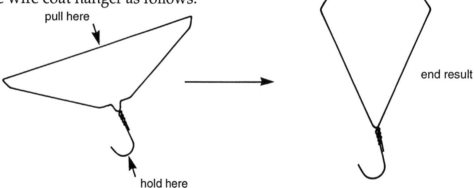

pull here

hold here

end result

2. Place the coat hanger contraption on your finger, and place a penny on the end of the wire.

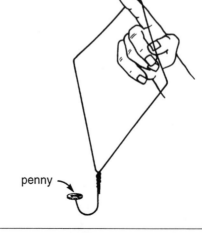

penny

3. While holding the coat hanger motionless with the penny in place, ask students to predict what will happen to the penny when you begin to twirl the contraption around your finger. Have students write their predictions under 1 on the student page.

4. Now start to swing the contraption slowly back and forth. Then progress to full loops. To successfully accomplish this, you will probably need to practice ahead of time.

5. As the contraption twirls around your finger, the penny should remain firmly in place on the wire.

6. Finish by swinging the contraption slower and slower so the penny eventually comes off. Try and catch the penny for effect.

Safety Concerns

In case something goes wrong and the penny flies off or the coat hanger goes flying, you should wear safety goggles and position students a safe distance away.

Outcomes and Explanations

The spinning produces a centrifugal force outward. This force holds the penny in place. The faster the spin, the greater the force. When the spin slows down, the force lessens until the penny falls. Discuss centrifugal force with students and have them write the explanation under 2 on the student page.

Application

Ask students where they may have seen or experienced such circular movements and forces. Examples of such movements and forces are found in certain carnival rides or in something as mundane as the clothes washer, when the drum filled with clothes and water quickly spins to separate the water from the clothes.

Take Home

Encourage students to try this activity at home and to make up other experiments using centrifugal force. Caution them to wear eye protection and keep spectators back a safe distance.

35 Human Tricks

Problem: *Can your body always do what you think it can?*

Can You Jump?

Stand with your back to the wall, touching it with your head, hips, and heels. Now try to jump up.

Can you jump up? Why or why not?

Can You Control Your Arms?

Stand in a doorway with the backs of your hands pressed against the door frame. Push outward with your arms as hard as you can for 60 seconds. Be sure to push the full amount of time. At the end of that time, relax your arms and turn sideways.

What did your arms do? Why?

Can You Lift Your Foot?

Stand with your right cheek, shoulder, side, and heel tight against a wall. Now try to lift your left foot off the floor.

Can you lift your left foot off the floor? Why or why not?

Can You Fold the Paper?

Try to fold a sheet of paper in half 10 times.

Can you fold the paper in half 10 times? Why or why not?

Can You Drop the Coin?

Fold your hands so they look like the picture. Have someone place a coin between the fourth digits of your hands. Now, try to open your fingers and drop the coin. You may not slide your fingers apart.

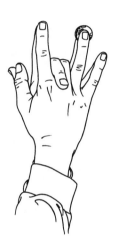

Can you drop the coin? Why or why not?

Can You Make the Points Meet?

Take a sharp pencil in each hand. Hold your hands about two feet apart with the pencil points facing each other. Close one eye and keep it closed. Now, bring your hands together and make the pencil points touch.

Could you bring the pencil points together? Why or why not?

Can You See Through Your Hand?

Roll a piece of paper into a tube and tape it. Take the tube in your right hand and hold it up to your right eye so that you can see through it. Now, raise your left hand, palm facing towards you, until it is a few inches in front of your left eye, with your little finger touching the side of the tube.

Can you see through your left hand? Why or why not?

35 For the Teacher

Objective

The tricks in this activity will test your students' powers of observation and critical thinking and their knowledge of their own bodies.

Materials Needed

- Arms?—one doorway for every two or three students
- Jump?—wall space
- Foot?—wall space
- Paper?—one sheet of newspaper per student
- Coin?—one coin per two students
- Points?—two sharp pencils per two students
- Hand?—one sheet of 8 1/2" × 11" paper and a small piece of tape per student

Curiosity Hook

Stand on one foot or jump up and down as students enter the classroom.

Setup

The setup for each trick is given on the student pages.

Safety Concerns

Use common sense, especially with pencil points.

Outcomes and Explanations

Jump?

To jump, the body must slightly bend, crouch, and coil. It is not possible to assume the proper body position to jump with heels, hips, and head against the wall.

Control Your Arms?

The signal from the brain to the arm muscles to contract (push) is sent so strongly and for so long that when the student stops pushing, the arm muscles continue to contract. Without the door frame to prevent movement of the arms, the arms float up as if controlled by someone else.

Lift Foot?

To lift one foot, you have to shift your center of gravity away from the support base. With half the body pressed against the wall, you cannot shift anything and, thus, cannot lift your foot.

Fold Paper?

Use math to see why it is not possible to fold the paper this many times. The first time you fold the paper you get 2 sheets; the second time, 4 sheets; the fifth time, 32 sheets. If you could manage to fold the paper the eighth and ninth times, you would be folding 512 sheets!

Drop Coin?

You can't pull your fourth digits apart because they are the weakest ones. They depend on the other fingers for leverage, so when you immobilize the other fingers, the fourth digits are helpless.

Points Meet?

You need both eyes open for depth perception. Using targets as small as pencil points and only one eye, the margin for error (missing) is enormous.

See Through Hand?

The right eye sees the inside of the tube, and the left eye sees the palm of the hand. In normal vision, the impressions received by each eye are combined to give a composite image in the brain. Because you are concentrating on looking through the tube in this trick, the image from the tube is transferred to the palm of the hand.

Application

The above activities reinforce the students' skill of observation to determine why an action works or doesn't work. Many actions that we take for granted are not always as simple as they seem and require the coordination of body and brain to achieve. Have students try to devise some tricks of their own using physical motion, balance, or visual perception.

Take Home

Encourage students to take these tricks from the classroom and use them to teach and amaze their parents, siblings, and friends.

36 Can You Light the Bulb?

Problem: *How many ways can you think of to light the bulb?*

Predict

1. You will work in groups on this activity. On the diagram below, two wires are used to light the bulb. Which designs will work? Test each design.

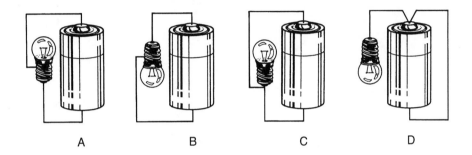

A   B   C   D

Conclude

2. Which designs worked and why? Why did some designs not work?

Predict

3. You will work in groups on this activity. On the diagram below, one wire is used to light the bulb. Which designs will work? Test each design.

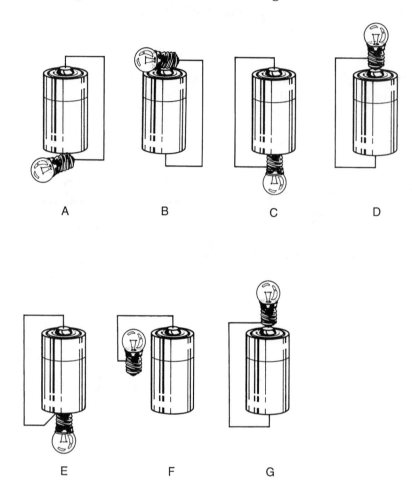

Conclude

2. Which designs worked and why? Why did some designs not work?

36 For the Teacher

Objective

In this activity, students will use their powers of critical thinking and problem solving to complete the circuit and light the bulb.

Materials Needed

- 1 flashlight battery (dry cell) per group of two or three students

- 1 small bulb (1.5 volt) per group of two or three students

- 2 wires per group of two or three students
  (Bare paper clips [not the plastic-coated type] or aluminum foil strips will also work.)

This activity requires a large quantity of materials. I borrow these materials from my physical science/physics department. If you do not have access to enough materials, you can have the students predict which designs will work and then demonstrate each one for them.

Curiosity Hook

Have the room dark when students enter. Switch on the lights and challenge students to test ways to switch on the lights they will be given.

Setup

For the first activity, each student team should have a battery, a bulb, and two wires. Using the diagrams on the student page, students should test which setups work to light the bulb. If you choose to do so, you can begin the activity by testing the first design for the students.

For the second activity, each student team should have a battery, a bulb, and one wire. Using the diagrams on the student page, students should test which setups work to light the bulb.

Safety Concerns

Use common sense.

Outcomes and Explanations

As long as one pole of the bulb is connected to either the + or – pole of the battery and the other pole of the bulb is connected to the other pole of the battery, a circuit is complete and the bulb will light (see illustration on next page). In the first activity, diagram

D does not form a complete circuit. In the second activity, diagram F does not form a complete circuit.

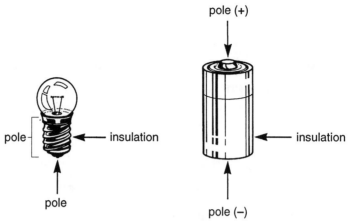

Application

Working in teams, have students think of and test other possible ways to light the bulb using two wires and then only one wire. Have them diagram the designs that work.

Creative Challenges

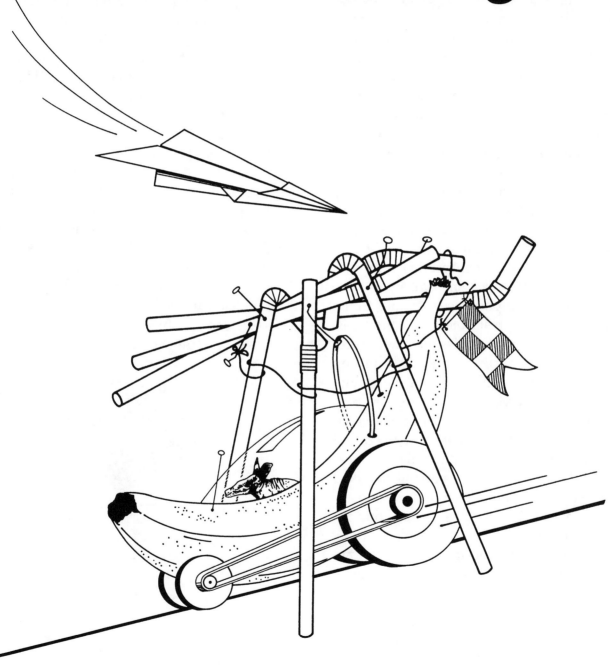

Creative Challenges Introduction

In each Creative Challenge, the teacher will present a problem (challenge) for you to solve. The teacher will give you a set of guidelines to follow for each challenge. Working within those guidelines, it will be up to you to meet the challenge and solve the problem. As you design and develop a solution to each problem, practice applying the science process skills demonstrated by the teacher in the Dynamo Demos: observing, predicting, experimenting, eliminating, and drawing conclusions. You may not use all of these steps in each challenge, but the basic process will remain the same.

In each challenge you will begin with a specific problem to solve. The basic materials for the project will be provided by your teacher, but in some of the challenges you will be allowed to add materials of your own. You will first observe the materials given and then predict how you might solve the problem. Once you have predicted some possible outcomes to the problem, you can start testing your ideas. In the challenge activities, you will be responsible for setting up and designing your own experiment within the guidelines given by your teacher. As you construct your design, you will need to analyze the different variables involved and eliminate alternatives until you have a reasonable solution. Finally, you will be asked to test your design. You will draw conclusions about the effectiveness of your design and other students' designs.

Suppose you had to solve the following problem: Do aardvarks eat chocolate pudding? If the teacher gave each student an aardvark and chocolate pudding, you would each need to set up an experiment to solve the problem. You would have to observe the aardvark to determine his eating habits and the types of food he normally eats. You would have to determine how to feed the pudding to the aardvark, how often to feed it to him, in what amounts to feed it, etc. Once you established your guidelines for feeding the aardvark the pudding, you would have to try out your experiment to see how it worked. If you were able to get the aardvark to eat the pudding, you would then need to record the results and compare your results with those of your classmates to draw some conclusions.

Of course, your teacher isn't going to give you an aardvark to feed, but we have included one in each challenge as a reminder to you to apply the basic steps of the scientific process as you attempt to solve each problem. Be observant, be creative, and, most of all, have fun!

1 Zippy Clippy

Challenge

Invent a new use for a paper clip.

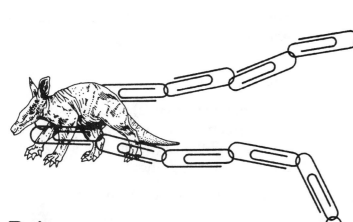

Rules

1. One large paper clip will be supplied by your teacher.

2. Your new invention must have some use other than clipping papers together.

3. The paper clip may be bent or cut in any way you want.

4. The paper clip by itself may be the whole invention or the paper clip could be used with other things, becoming part of a larger invention. The paper clip can be attached to other things or other things can be attached to the paper clip.

5. Brainstorm your design then make a detailed labeled drawing of your invention. Think SAFETY.

6. Once you have your invention designed and drawn, explain it to your teacher. If your teacher decides that your design is safe and practical, you may go ahead and begin building.

7. Once everyone has completed his/her invention, you will explain your invention to the rest of the class and demonstrate how it works.

1 For the Teacher

The Challenge

Inventors are creative and observant people who see connections, patterns, and solutions that others often overlook. This activity tries to fan the spark of invention in your students by challenging them to dream up a new function for a common object—a paper clip.

The Rules

1. Have students work alone or in teams.

2. You will need to supply one paper clip per student or per team. Large metal clips that can be bent or cut are preferable.

3. Depending on your time restrictions, let students either work on this activity during class time or have them brainstorm, design, and build their inventions outside of class but present and demonstrate them to the rest of the students during class time.

4. After students have had time to brainstorm and come up with a design for some sort of invention, require them to submit a labeled diagram and explanation of the invention before you give them clearance to begin construction. Think safety and practicality.

5. Once construction is complete, have students use their labeled diagrams to explain their inventions to the rest of the class. Then, if practical, have them demonstrate how their inventions actually work.

Safety Tips

Discourage designs that have sharp pointed projections and avoid those that employ flying projectiles.

Stumped?

If you have a student(s) who gets stuck for a paper clip invention idea, offer the following slightly silly suggestions to get the creative process going:

- 1/2 of a pair of metal chopsticks
- a backscratcher for a short person
- a small lightning rod
- a roasting stick for cocktail weiners

Awards and Recognition

1. Display students' design diagrams and actual inventions where other students, teachers, and parents can appreciate them.

2 Wild Blue Yonder

Challenge

Keep a single piece of paper in the air for as long as possible without touching it.

Rules

1. You will be given two sheets of paper both the same size.

2. Your "airplane" must be made from only one sheet of paper or the pieces cut from one sheet of paper. Use one sheet of paper for practice and the other for the actual competition.

3. The sheet of paper may be folded or cut in any way you choose.

4. Tape, staples, and/or paper clips may be used to hold pieces of your "airplane" together. Nothing else may be attached to your plane. Remember—the heavier the design, the faster it will fall.

5. Only one "airplane" per person or group may be entered.

6. Your plane must be built within the time limit set by the teacher.

7. Each plane will be hand launched but cannot be touched or aided in any way after launch.

8. Time will start once the plane leaves your hand and stop once the plane hits the floor or some object.

9. Each person or group will be allowed three trial runs. Only the best of the three trial runs will be counted.

2 For the Teacher

The Challenge

In this activity, you will see the many creative methods students use to keep a piece of paper aloft as long as possible.

The Rules

1. Have students work alone or in teams.

2. Provide each student or team with two sheets of paper both the same size. The size of the paper is up to you. Standard 8.5 × 11-inch paper works well.

3. You will also need to provide tape, a stapler, and paper clips in case students or teams wish to incorporate them into their designs.

4. Consider this—I once saw a group of students at a science fair win a competition similar to this by merely dropping a flat sheet of paper. Technically, these students followed the letter of the rules, but did they embrace the intent of the rules? Something to consider beforehand.

5. Set a time limit. I allow my students to work on their designs at school during class time. I set aside several periods for design, construction, and testing and one period for competition.

6. Hold the competition indoors in a large area like a gymnasium where the air is calm.

7. Each student, or someone from each team, must hand launch the plane—assisted takeoffs using rubber bands or model rocket engines are not allowed.

8. A hand-held stopwatch is satisfactory for timing. Start timing when the plane is released and stop timing when the plane hits the floor or some object.

9. Give each student or team three trial runs but count only the best time of the three attempts.

Safety Tips

None

Stumped?

If available, books on making paper airplanes might serve as inspiration.

Awards and Recognition

1. Display a chart showing the time aloft for each plane.

3 Towers in the Sky

Challenge

Construct the tallest structure possible using only newspaper and tape.

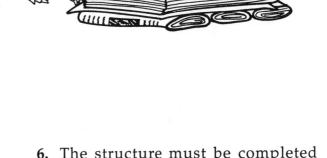

Rules

1. The teacher will provide you with newspaper.

2. You will get only three feet (36 inches) of tape. Use it wisely!

3. Your structure cannot be attached to anything other than the floor, and it cannot lean against anything.

4. The structure must stand on its own long enough to be measured. You cannot hold up or support your structure while it is being measured.

5. The height of the structure will be measured from the floor to the highest point on the structure.

6. The structure must be completed within the time limit set by the teacher.

3 For the Teacher

The Challenge

In this challenge, students work against the clock using limited materials to construct a tower.

The Rules

1. Let students work alone or in design teams. I find it more convenient to have students work in teams of two or three.

2. You will need to supply each student or team with newspaper. I let my students or teams have all the paper they need, but you could limit each student or team to a fixed number of pages of newspaper. Either way, you need to plan ahead and have plenty of newspaper on hand.

3. You will need to supply each student or team with three feet of tape. One-inch masking tape seems to work the best.

4. Students often build towers that can go higher than most classroom ceilings will allow. Therefore, conduct this activity in an area like a gymnasium with adequate overhead clearance.

5. Towers may grow higher than students can reach, so you will also need to supply stepladders. Sturdy 8-foot or 10-foot stepladders work quite nicely. It may not be possible or practical to provide ladders for each student or even for each team. However, since they are working against the clock, students grow understandably frustrated if they have to wait long for access to a stepladder. One ladder per three or four students or one ladder per two teams is a workable minimum.

6. Constantly remind students that the base is the key. Only towers built with a sound, stable base will achieve success.

7. The base of the tower may be taped to the floor. However, no other part of the tower may be attached to anything, nor may the tower lean against anything for support.

8. The tower must stand without student assistance long enough to be measured. Some towers will stand but droop. How will you measure these? Will you measure from the floor to the highest vertical point on the tower or will you measure the total length of the tower from base to tip disregarding droop? You need to decide this beforehand and inform students before the competition begins.

9. A 16-foot metal tape should be adequate for measuring tower heights.

10. Set a time limit. I give my students one class period to construct their towers. I issue the challenge, let the students pick their design teams, and I prepare the building supplies, except for the tape, the day before. The next day as students come in, they get their building supplies and tape and begin construction. Leave enough time at the end of the period for final measurements and cleanup.

Safety Tips

Urge caution when working from stepladders and NEVER permit a student to stand on the top step of a stepladder.

Awards and Recognition

1. Display a chart showing the height of each student's or team's tower.

4 A Heavy Load

Challenge

Design and build a support that will hold as much weight as possible.

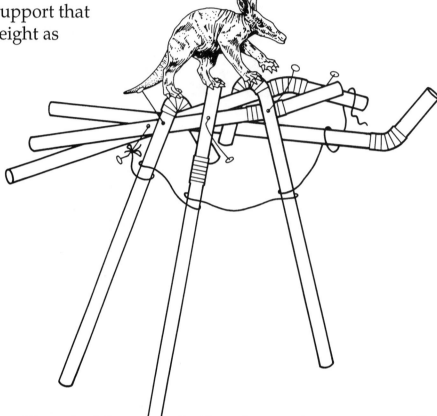

Rules

1. Each student or team will get the following construction materials:

   - 20 plastic drinking straws

   - a measured lump of modeling clay

   - 20 straight pins

   - 2 feet (24 inches) of tape

   - 2 feet (24 inches) of string

   - scissors for cutting straws, tape, or string

2. Only the straws, clay, pins, string, and tape may be used to construct the support. The scissors may not be part of the support.

3. You do not have to use all the materials.

4. The straws may be cut to any length.

5. The minimum height of the overall structure must be six inches.

6. The support must be flat enough on top to hold the weighing device.

7. The support must be completed within the time limit set by the teacher.

8. The support that holds the most weight before buckling will be declared the winning design.

4 For the Teacher

The Challenge

In this challenge, students work against the clock, using limited materials to construct a weight-bearing support.

The Rules

1. Let students work alone or in teams. From a time and materials standpoint, I find it more convenient to have students work in teams of two or three.

2. Give each student or team a lump of clay weighing approximately 50 grams. It is not essential that each student or team get precisely the same weight of clay. You could just eyeball it and give each student or team a lump of clay roughly the size of a Ping Pong ball.

3. To keep clay off tables or desks, have students build their supports on newspaper or use wax paper if you wish to try and salvage some of the clay when the activity is over.

4. The supports must be built in such a way that you can stack some kind of weight on them. I require my students to build supports with a flat top. When it comes time to add weights, I place a 10-inch × 10-inch piece of thin plywood on the support. I then add weights borrowed from the athletic weight room until the support eventually buckles. Another but less accurate method would be to stack books of approximately the same size (like a set of encyclopedias) on the support until it collapses.

5. Set a time limit. I give my students one class period to construct their supports. I issue the challenge, let the students pick their design teams, and I prepare the building supplies, except for the tape, the day before. The next day as students come in, they get their building supplies and tape and begin construction. Leave enough time at the end of the period for final measurements and cleanup.

Safety Tips

Urge students to practice safe handling of pins and scissors.

Awards and Recognition

1. Display a chart showing the weight (or number of books) held by each student's or team's support.

5 Indybanapolis 500

Challenge

Design a contraption to carry a banana as far as possible.

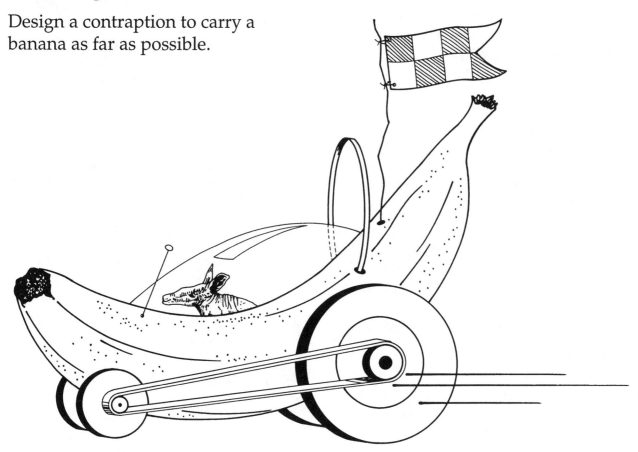

Rules

1. Your bananamobile must be powered only by two rubber bands. The teacher will supply the rubber bands.

2. The banana must remain in one piece.

3. Things may be attached to or stuck into the banana, or the banana may be attached to other objects.

4. Your bananamobile must be built within the time limit set by the teacher.

5. You will be given several trial runs. You can decide which run counts.

6. The bananamobile that travels the greatest distance will be declared the winning design.

5 For the Teacher

The Challenge

In this activity, students are challenged to see connections with and make combinations of objects that are not normally associated with each other.

The Rules

1. Have students work alone or in teams. From a time and materials standpoint, I find it more convenient to have students work in teams of two.

2. You will need to supply one banana for each student or team. To increase longevity, use bananas that are as green as possible.

3. You will also need to supply two rubber bands for each student or team. To ensure fairness, each student or team should have the same size rubber bands. I use rubber bands that are 1/4-inch wide by 4 inches long.

4. You can have students work on their designs either during class or on their own outside of class. I present this challenge to my students, let them select their teams, and give them their banana and rubber bands three days before the competition. They work on their designs outside of class and then bring them to school the day of the competition. One class period is usually sufficient to allow everyone to run their designs.

5. Students usually take one of three approaches to this challenge. They will attach wheels to the banana itself to make it into a car, attach the banana to a toy car, or use the rubber bands to catapult the banana. I allow all these approaches. The real challenge is in designing a system to use the rubber bands as a power source.

6. Some of these contraptions may travel surprisingly far, but often they don't travel reliably straight. Running the bananamobiles down a hallway might work, but I suggest you run them in a large open area like a gymnasium.

7. If time permits, give each student or team several trial runs and allow them to pick the run they wish to count.

8. A 50-foot or 100-foot tape measure works well. You might be able to borrow such a tape measure from your industrial technology department, athletic department, or a local construction company.

Safety Tips

It is very difficult to turn rubber bands and a banana into something dangerous but, as usual, think safety and keep a close watch for any potentially harmful designs or situations.

Awards and Recognition

1. Display a chart showing the distance traveled by each student's or team's bananamobile.

2. Display any of the actual bananamobiles that remain in reasonably good condition after the competition. Limit this to a short period as the bananas deteriorate rapidly due to the rigors of competition.

6　Tug of War

Challenge

Construct the strongest link in a chain.

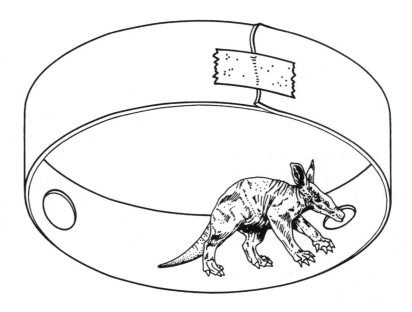

Rules

1. You will get the following materials from your teacher:

   • 1 piece of paper, 8 1/2 × 11 inches

   • 1 stick-on label or equal size piece of masking tape

   • 1 piece of string, 6 inches

2. Your chain link may be any size you wish.

3. Your chain link must have two holes in it large enough for a heavy cord to pass through.

4. Put your name on your link.

5. Your link must be constructed within the time limit set by the teacher.

6. Your chain link will be attached to other students' links by a heavy cord to form a chain.

7. The chain will be pulled in opposite directions. When a student's link tears, he/she is out of the competition.

8. The torn link(s) will be removed and the other links reattached. This tug of war will continue until only one link remains. The student designer of the strongest link will be declared champion.

6   For the Teacher

The Challenge

In this activity, students using limited materials work against the clock to construct a device that will resist tearing.

The Rules

1. Have each student work alone.

2. Provide each student with one piece of 8 1/2 × 11-inch paper, one stick-on label or equivalent-size piece of masking tape, and one 6-inch piece of string. The label or piece of tape should be at least 2 square inches. The 6-inch piece of string is to be part of the link and is not for attaching the links together.

3. Heavy cord will be needed. You will need enough pieces of cord to tie all the links together and several pieces to attach to each end of the completed chain.

4. Each student's link must have two holes in it to allow attachment to other links. A paper punch will make holes large enough to accommodate heavy cord.

5. Set a time limit. I require my students to complete this in one class period. Leave time to conduct the tug-of-war to decide the winner.

6. If many students are involved, begin with several chains until the number of contestants becomes manageable with one chain.

7. Attach the links together with heavy cord. Make sure each student has his/her name or some type of identifying mark on the link.

8. On both ends of the chain, attach a three- to four-foot piece of heavy cord for pulling.

9. Have students pull on each end of the chain. Have students pull <u>slowly</u> but firmly until one or more links tear. Remove the torn link(s), reattach the remaining links, and pull again. The last whole link is the winner.

Safety Tips

Don't let students get too carried away with the tug of war; stress pulling slowly.

Awards and Recognition

1. Display the winning link where it can be appreciated by students, teachers, and parents.

2. Award a paper-ring crown to the winner.

7 Span the Gap

Challenge

Design and construct a bridge of straws that will hold the most weight possible.

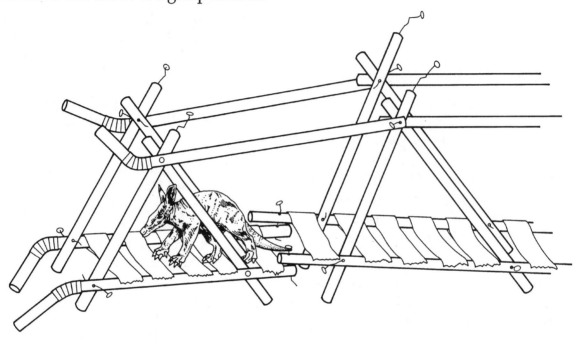

Rules

1. You will get the following materials from your teacher:

   - 12 inches of masking tape

   - 15 plastic straws

   - 12 straight pins

   - scissors as needed

2. The scissors are for cutting straws if necessary. The scissors cannot be part of the bridge.

3. The bridge must span 12 inches. Two tables or desks will be placed 12 inches apart, and the bridge must be built across this gap.

4. The bridge may not be taped to the tables or desks.

5. Your bridge must be built within the time limit set by the teacher.

6. You cannot touch or support your bridge in any way.

7. The bridge that holds the most weight will be declared the winning design.

7 For the Teacher

The Challenge

In this activity, students are not only challenged to construct a bridge across a gap but also to make the bridge sturdy enough to hold weight.

The Rules

1. Have students work alone or in teams. From a time and materials standpoint, I find it convenient to have students work in teams of two or three.

2. You will need to supply each student or team with 12 inches of masking tape, 15 plastic straws, and 12 straight pins. Scissors should be made available for cutting straws.

3. Bridges must be constructed across a 12-inch span. Properly positioned tables or desks will provide the necessary gap.

4. Set a time limit. I give my students one class period to construct their bridges. I issue the challenge and let the students pick their design teams. I prepare the building supplies, except for the tape, the day before. The next day as students come in, they get their building supplies and tape and begin construction. Leave enough time at the end of the period for final measurements and cleanup.

5. The challenge from the teacher's standpoint is to figure out how to attach enough weight to the bridges to collapse them. I recommend that you construct a weight basket or container that can be attached (hung) from the center of each bridge. Your weight basket or container should be sturdy but as light-weight as possible. Some bridges may not even support an empty weight basket or container.

6. Consider borrowing several weight sets from the physical science department.

7. Add weights to the weight basket or container until the bridge collapses. Adding a few small weights at a time allows for greater accuracy of measurement (and it heightens the excitement).

8. The bridge that supports the most weight should be declared the winning design.

Safety Tips

Urge students to practice safe handling of pins and scissors.

Awards and Recognition

1. If possible, display the winning bridge.

2. Display a chart showing how much weight each bridge held.

8 Look Out Below!

Challenge

Construct a contraption to prevent a fresh egg from breaking when it is dropped.

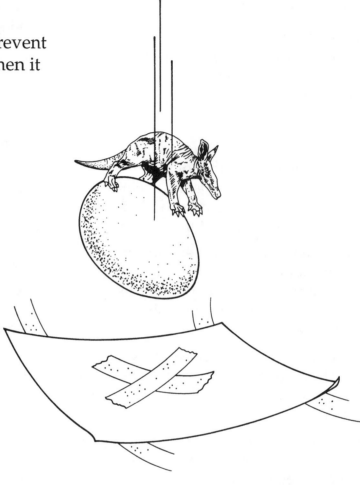

Rules

1. You will be given the following materials by your teacher:

   - 1 fresh egg
   - 10 sheets of paper
   - 5 feet (60 inches) of masking tape
   - scissors as needed to cut paper and/or tape

2. Only the paper and tape may be part of the contraption, but you don't have to use all the paper and tape.

3. Your contraption, with the egg placed in it, will be dropped from a height determined by the teacher. Remember—it's not the fall that breaks the egg, it's the sudden stop.

4. Your contraption must be constructed within the time limit set by the teacher.

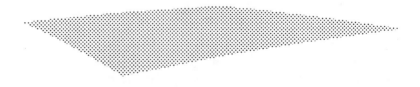

8 | For the Teacher

The Challenge

Students view eggs as fragile objects and rightfully so. In this activity, they are challenged to use minimal materials to build a device that will protect and cushion a fresh egg when the egg is placed inside the device and dropped onto a hard surface.

The Rules

1. Have students work alone or in teams.

2. You will need to supply one fresh egg, 10 sheets of paper (8 1/2 inch × 11 inch is a challenging size), and 5 feet of masking tape per student or team. Make scissors available to cut paper or tape as needed. The scissors cannot be part of the device.

3. Set a time limit. I give my students two class periods. They build their devices during one class period and then we drop them the next day during class. If you do this entire activity during one class period, leave enough time at the end to drop all the devices.

4. To make this activity especially challenging, drop the devices from the greatest practical height onto as hard a surface as possible.

Safety Tips

1. Be aware of the risks of having students up on ladders or crawling on top of the school to drop their devices.

2. Make sure students stand clear of falling devices. If a student gets injured, legally the yolk could be on you.

Awards and Recognition

1. It will not be practical to display student devices since they will most likely have to be torn apart to determine whether the egg survived the fall. And if the egg doesn't survive, the devices will be soggy and dripping gunk.

2. Display a chart showing which students' or teams' eggs did survive.

3. Award gag gifts, such as chocolate-covered eggs, as trophies to all participants.

9 The Hot-Air Express

Challenge

Design and construct a car that will go
the greatest possible distance powered
only by a balloon.

Rules

1. You will be given the following materials by your teacher:

   - 1 hamburger carton

   - 4 Styrofoam cups

   - 4 soda straws

   - 2 feet (24 inches) of masking tape

   - a balloon

   - scissors as needed for cutting materials

   - glue as needed for attaching materials

2. The scissors and glue bottle may not be part of the car.

3. The car must be constructed within the time limit set by the teacher.

4. Each student or design team may run only one car.

5. Depending on time available, each car will get several trial runs. The best run will be counted.

6. The car that travels the greatest distance will be declared the winning design.

9 For the Teacher

The Challenge

In this activity, students are challenged to work within a time limit using minimal materials to construct a car powered only by a balloon.

The Rules

1. Have students work alone or in design teams. I find it more convenient to have my students work in teams of two or three.

2. You will need to supply the following for each student or team:

   - 1 hamburger container
     (Local restaurants will often donate these.)

   - 4 Styrofoam cups
     (The 6-ounce size works best.)

   - 4 soda straws
     (The jumbo size works best.)

   - 2 feet of masking tape
     (The 1/2" size works best.)

   - a balloon
     (The 12" round size works best.)

   - scissors and white glue as needed

3. Set a time limit. I give my students two class periods. They build their cars during one class period and then we run them the next day during the class period. If you do this entire activity during one class period, leave enough time at the end to run all the cars.

4. The cars seldom travel in straight lines, so conduct the trial runs in a large area such as a gymnasium.

5. Give each car as many trial runs as practical time-wise. Count only the best run for each car.

6. For determining the distance the cars travel, a 20-foot measuring tape should be adequate but a 50-foot measuring tape would be best.

7. On the floor, put a piece of tape with a mark down the middle for a starting line. Measure distance from the mark on the tape to the front end of the car.

Safety Tips

Urge students to practice safe handling of scissors.

Awards and Recognition

1. Display the cars where they can be appreciated by other students, teachers, and parents.

2. Display a chart showing how far each car traveled.

10   Mail It!

Challenge

Using a single rubber band, design and build a contraption to send a written message at least fifteen feet.

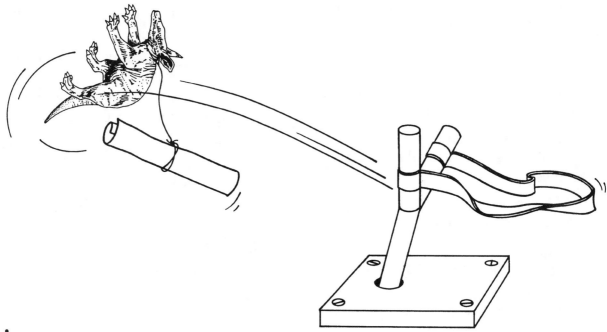

Rules

1. The teacher will supply you with one large rubber band. You must supply everything else your design needs.

2. The only power source must be a single rubber band. No springs, motors, or rocket engines are allowed.

3. The message may be rolled, wheeled, bounced, shot, or whatever. The delivery method is up to you.

4. The message must be launched from a surface or the floor. It cannot be launched from your hands.

5. The message to be delivered must be written on a piece of paper. The size of the paper and what the message says are up to you.

6. If necessary, and if time allows, you will get several trial runs.

7. Your message delivery system must be completed within the time limit set by the teacher.

10 For the Teacher

The Challenge

In this activity, students are given a single rubber band and challenged to use this rubber band to send a written message a minimum distance.

The Rules

1. Have students work alone or in teams.

2. You will need to supply one large rubber band per student or per team.

3. Depending on your time restrictions, let students either work on this activity during class time or have them brainstorm, design, and build their delivery systems outside of class but test their inventions during class time.

4. The messages may go all over the place, so I recommend you test the systems in a large open area like a gymnasium.

5. Place a piece of tape on the floor (start line) and then fifteen feet away place another piece of tape (finish line). The messages must be launched from the starting line on the floor or another flat surface.

6. If necessary, and if time permits, give each student or team several trial runs.

7. You could make this activity even more difficult by challenging students to deliver their messages as close as possible to exactly fifteen feet.

Safety Tips

Caution students not to shoot rubber bands or launch other objects with the rubber band contraption.

Stumped?

If you have a student(s) who gets stuck for an invention idea, brainstorm sports terms like racket, bat, throw, shoot, pitch, and so on. Then ask questions like How could you "pitch" a message? or How could you "shoot" a message?

Awards and Recognition

1. Display each student's or team's message delivery system.

2. Display a chart showing which systems met the challenge.

11 Go Slow Ramp Race

Challenge

Construct a series of ramps for a marble to get from the top of a shoe box to the bottom in the longest time possible.

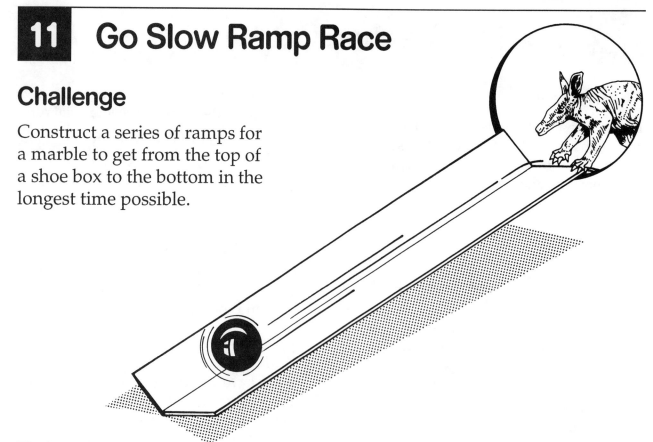

Rules

1. The teacher will provide you with the following materials:

   • a shoe box

   • heavy paper and thin cardboard for building ramps

   • glue and tape

   • scissors for cutting and shaping ramps

   • a marble for practice runs

2. The ramp system must start in a top corner of the box.

3. A marble-sized entrance hole must be made in the top corner of the box directly above where the ramp system starts.

4. The ramp system must be constructed of only heavy paper and/or thin cardboard glued or taped together.

5. The lid of the box must come off so the inside can be examined.

6. A marble-sized exit hole must be made in the lower corner of the box where the ramp system ends.

7. Your ramp system must be built within the time limit set by the teacher.

8. Timing will start when the marble is dropped in the entrance hole at the top of the box and timing will stop when the marble rolls out the exit hole at the bottom of the box. The system that makes the marble take the longest time to exit will be declared the winning design.

9. If time permits, you will be given several trial runs. You may make adjustments to your ramp system between trial runs. Only the longest run will count.

11   For the Teacher

The Challenge

In this activity, students will win the race if their marble is the slowest instead of the fastest.

The Rules

1. Have students work alone or in teams. I find it more convenient to have my students work in teams of two or three.

2. You will need to supply the following items for each student or team:

   - a shoe box
     (Ask students to bring these from home or ask local shoe stores to donate some.)

   - paper and cardboard of assorted thicknesses for constructing ramps

   - white glue and tape

   - scissors

   - a marble for practice runs

3. Depending on your time restrictions, let students either work on this activity during class time or have them brainstorm, design, and build their delivery systems outside of class but test their inventions during class time.

4. A hand-held stopwatch will work for timing. Start timing once the marble is dropped into the entrance hole and stop once it comes out the exit hole.

5. If time permits, give each student or group several trial runs. Allow them to adjust or alter their ramp system between trial runs. Count only the longest trial run for each student or team.

Safety Tips

Urge students to practice proper safety procedures when working with scissors.

Awards and Recognition

1. Display each student's or team's box with ramp system.

2. Display a chart showing the times for each student's or team's system.

12 | Cannon Shots

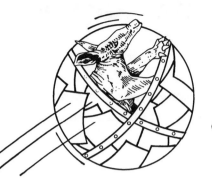

Challenge

Design and construct a mousetrap contraption to launch an object towards a target.

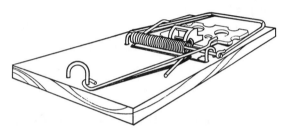

Rules

1. The teacher will provide you with one mousetrap. You must supply everything else you need to build your launch system.

2. The mousetrap may be modified in any way you choose, but the only energy source used to launch your object must be the spring of the mousetrap.

3. You must also design and build the object to be launched. There is no size or weight limit on the object to be launched. You must supply everything needed to build your launch object. Make your launch object as brightly colored as possible so it will be easy to see.

4. You must launch your object from the floor.

5. Your launcher must be built within the time limit set by the teacher.

6. You will receive points for accuracy in hitting a target. The scoring system will be as follows:

    *   in the target   = 10 points
    *   6 inches or less   = 9 points
    *   12 inches or less   = 8 points
    *   18 inches or less   = 7 points
    *   24 inches or less   = 6 points
    *   30 inches or less   = 5 points
    *   36 inches or less   = 4 points
    *   over 36 inches   = no points

7. The distance from the target will be determined by where your object first hits, not where it comes to rest.

8. You will be given several trial runs; points for each run will be recorded. The launch system with the highest total number of points will be declared the winning design.

12 | For the Teacher

The Challenge

In this activity, students are challenged to use the spring of a mousetrap to launch an object of their own design as close to a target as possible.

The Rules

1. Have students work alone or in teams. I find it more convenient to have my students work in teams of two or three.

2. You will need to supply one mousetrap per student or team.

3. Allow the students to modify the mousetrap as they wish, but the only power source for launching their projectiles must be the spring of the mousetrap. Any other materials needed to build the launch system should be provided by the students.

4. Require the students to also design and build the object to be launched. I suggest you place a monetary limit on how much can be spent on materials to build a launch system and launch object.

5. Depending on your time restrictions, let students either work on this activity during class time or have them brainstorm, design, and build their launch systems and launch objects outside of class but test their inventions during class time.

6. Tape several large sheets of paper together and draw a target on them. Make the center of the target a 10-inch circle (or 8-inch if you want it to be more challenging). Then draw circles around the target center every 6 inches outward to 36 inches.

7. You can decide how far away to place the target. I find 15–20 feet to be a challenging distance. The main thing is that students know how far away the target will be before they begin building their systems.

8. The launch objects fly all over the place, so I suggest you do this activity in a large open area like a gymnasium.

9. Place a piece of tape on the floor and then position the target an appropriate distance away. All launches must take place on the floor at or behind the tape.

10. A yardstick will suffice for a measuring device. Measure from where the object first hits, not from where it comes to rest. You will need to enlist the aid of students as spotters to determine precisely where the objects hit.

11. Use the distance-point system outlined on the student page of this activity to keep score. Try to give each student or team at least three trial runs and then add up their scores. High score is the winning design (or luckiest launches).

Safety Tips

1. You may get some pinched fingers from mousetrap springs.

2. Require students to verify the safety of their launch systems and launch objects before they begin construction. Do not allow students to build projectiles that could cause injury if they were to strike someone.

13 Pasta Planks

Challenge

Build an arm of uncooked spaghetti that extends as far as possible before it touches the floor.

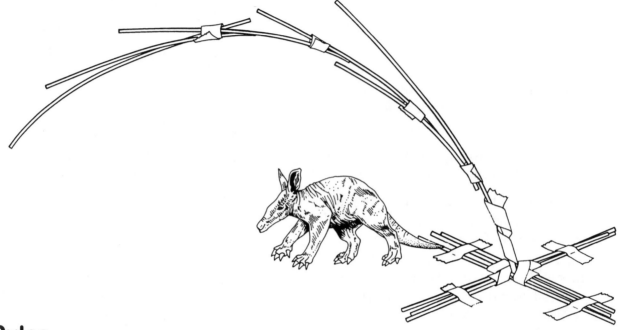

Rules

1. The teacher will supply the following materials:

   • as much uncooked spaghetti as needed

   • 24 inches of masking tape

2. The base (beginning) of your arm must be attached to the edge of the tabletop.

3. Using pieces of uncooked spaghetti (pasta planks) and tape, build an arm that reaches out from the base as far as possible without touching the floor.

4. You may break the spaghetti into smaller pieces.

5. If your arm touches the floor, you may break it back until it does not touch the floor. You will be allowed to support and straighten your arm for measurement.

6. Your arm will be measured from the base outward to the end of the arm.

7. Your arm must be built within the time limit set by the teacher.

8. The longest arm will be declared the winning design.

13 | For the Teacher

The Challenge

In this activity, students are challenged to build within a set time a spaghetti cantilever using minimal materials. (A cantilever is a horizontal arm which extends beyond its point of support.)

The Rules

1. Have students work alone or in teams. I find it more convenient to have my students work in teams of two or three.

2. You will need to supply the following for each student or team:

   • uncooked spaghetti
   (One package of uncooked spaghetti per several students or per team should be sufficient.)

   • 24 inches of masking tape

3. Set a time limit. I give my students one class period to construct their arms. I issue the challenge, let the students pick their design teams, and prepare the building supplies (purchase the spaghetti), except for the tape, the day before. The next day as students come in, they get some spaghetti and tape and begin construction. Leave enough time at the end of the period for final measurements and cleanup.

4. Have students build their arms on the edge of a table or desk. The base of the arm should be taped to the top of the table or desk. Caution students to use a minimal amount of their tape attaching the base; save as much tape as possible for the arm and braces. Let students decide how far back from the edge of the table or desk to start building the base of their arm.

5. Measure arms from the base to the end of the arm. A measuring tape will be needed to determine the length of each arm. A 10-foot tape should be sufficient.

6. Measure only arms that do not touch the floor. Allow students to break off the end of their arms if necessary until they do not touch the floor. Also allow students to support and straighten their arms only during measurement.

7. The longest arm that doesn't touch the floor should be declared the winning design.

Safety Tips

Use common sense.

Awards and Recognition

1. It really isn't practical to save and display the arms themselves but do display a chart showing the length of each arm where it may be appreciated by other students and teachers and by parents.

14 Will the Yolk Be on You?

Challenge

Which pair of students can toss a fresh egg the farthest without breaking it?

Rules

1. The teacher will supply the following materials:

   - 1 fresh egg
   - safety goggles
   - protective clothing

2. Work in teams of two.

3. Each team will toss the uncooked egg back and forth. If the egg remains unbroken, the distance between team members will be increased and the egg tossed again. The winning team will be the one that can toss the egg the greatest distance back and forth without breaking the egg.

4. Objects resist any change in motion. Motionless objects remain still while objects in motion keep moving. This property of objects is called inertia. Energy is required to start motionless objects moving (acceleration) and energy is required to stop moving objects (deceleration). Acceleration is what pushes you back in the seat of an amusement park ride that moves rapidly forward. In this situation, you are an object at rest and you resist moving. Deceleration is what can throw you into the dashboard in a car accident if you are not wearing a seat belt. In this situation, you are an object in motion and you resist stopping.

5. Remember, it's not the speed (acceleration) that breaks the egg, it's the sudden stop (deceleration). If the force of stopping is greater than the strength of the egg shell, the egg will break. The challenge is to catch the egg in such a way as to make the force of stopping as small as possible. Of course, as the distance increases between you and your teammate, this becomes harder and harder to do because the egg will be moving faster and faster.

14 For the Teacher

The Challenge

In this challenge students try to delay the inevitable consequences of inertia as long as possible.

The Rules

1. Have students work in teams of two. In case of an odd number of students, someone can be in two groups.

2. You will need to supply the following for each team:

   - 1 fresh egg
     (Balloons filled with water could be substituted for the eggs.)

   - safety goggles and lab aprons
     (These may be borrowed from your secondary physics or chemistry departments.)

3. Make sure all students understand the information in student rule 4 that explains this challenge.

4. This activity is best conducted outside. Depending on space available, teams can compete one at a time or all together. You may want to have pre-measured throwing lines set up ahead of time.

5. Start team members 15 feet apart. A catch is complete when both team members catch the egg without breaking it.

6. After each complete catch, increase the distance between team members in 5-foot increments until the egg breaks.

7. The team that can toss the egg the greatest distance back and forth should be declared the Champion Deceleration Force Fighters.

Safety Tips

Team members should wear safety goggles and lab aprons when tossing the egg. Other students should stand well back from the action.

Awards and Recognition

1. Display a chart showing the distance each team tossed their egg.

2. Award appropriate gag gifts, such as candy eggs, to all participants.

15 Bean Stalks to the Sky

Challenge

Grow the tallest plant possible within a set time.

Rules

1. The teacher will give you 5 seeds. You must supply everything else.

2. You will need to make the following decisions about your seeds:

   - what material will you plant your seeds in—soil, sand, etc.?
   - what kind of container(s) will you plant your seeds in?
   - how many seeds will you plant in each container?
   - how deep will you plant the seeds?
   - how much light will the plants receive?
   - how much and how often will you water the plants?
   - will you fertilize the plants?
   - if you will fertilize, how much and how often?
   - and, most important, how can you adjust and regulate the seeds and/ or the conditions the seeds are in to get the tallest plant in the shortest amount of time?

3. You should keep a record of the conditions under which you grow your plant, and any changes you make.

4. Your teacher will set a time limit on how long you have to grow your plants.

5. A word of caution! While plants need some water and fertilizer, go easy on these things. Your plants will suffer more from too much water and fertilizer than from too little.

6. At the end of the time limit set by the teacher, the plants will be measured. The longest plant, measured from the top of the soil in the container to the tip of the plant, will be declared the winner.

15   For the Teacher

The Challenge

In this activity, students delve into the biological realm by manipulating the conditions of plant growth to produce the tallest plant possible within a set time limit.

The Rules

1. Have students work alone or in teams.

2. Provide each student or team with at least 5 seeds. Lima beans, green beans, or pinto beans will work. Peas are also satisfactory, but dwarf types will not produce the longer stems of regular varieties. I prefer mung beans for their rapid growth and elongated stems. Check local suppliers or seed catalogs for mung beans, or order from one of the following sources:

   - Connecticut Valley Biological
     82 Valley Road
     P.O. Box 82
     Southampton, MA 01073

   - Carolina Biological Supply
     Company
     2700 York Road
     Burlington, NC 27215

   - Ward's Natural Science
     Establishment, Inc.
     P.O. Box 92912
     Rochester, NY 14692-9012

3. If appropriate and necessary, you may need to supply students or teams with planting medium and/or planting containers. If a student or team plan on using soil as their planting medium, commercial potting soil is best. Soil from a garden or road ditch may contain fungi that can cause rot-ting seeds or damping off (a fungal disease causing seedlings to wilt and die) of the seedlings.

4. Allow students to place their plant growth containers in the classroom or at home in places that they think will have maximum growth conditions.

5. Students should record the conditions under which they grow their plants, and record any changes they make.

6. You may face the situation where none of a student's or team's seeds germinate or if they do, they quickly die. You may choose to grow a few extra plants.

7. Set a reasonable time limit. At least several weeks may be necessary to get satisfactory results. Of course, the time limit can always be altered if necessary.

8. At the end of the time limit, measure each student's or team's plants. Measure from the top of the soil in the pot to the very tip of the plant. You can discuss the variables involved in growing the plants to determine which conditions caused the most growth. The longest plant should be declared the winner.

Safety Tips

1. Caution students about putting seeds in their mouths or allowing younger children in their household to do the same. Choking could result. Also, some seed companies coat seeds with a fungicide (usually a bright pink color) that could be harmful if ingested.

2. Urge students to handle fertilizers safely. Fertilizer containers should be

stored out of the reach of small children and fertilizer/water mixtures should not be left where they might be accidentally ingested by someone.

3. If students build plant growth chambers from flammable materials and place incandescent lights inside, there is a risk of fire.

Stumped?

1. Students might try manipulating light conditions. Up to a point, plants will grow longer and somewhat faster in total darkness. Placing cardboard tubes over plants so that light comes only in the top of the tube is a strategy I have also seen used with some success. If students have access to plant lights and timers, they might wish to try increasing the hours of light the plants receive.

2. Students might try manipulating temperature conditions. Making plant-growth chambers from cardboard boxes and placing incandescent bulbs inside to raise the temperature is one strategy. CAUTION: If students build these plant-growth chambers from flammable materials, there is a risk of fire. Incandescent bulbs get hot enough to set on fire any flammable materials they contact.

3. Students might try manipulating water/fertilizer conditions. Caution students to go easy on water and fertilizer. Students often have the mistaken idea that if a little water and fertilizer are good, a LOT of water and fertilizer will be even better. This is not true.

Awards and Recognition

1. Display a chart showing the length of each student's or team's plants.

2. You might consider giving the winning student or team a small house plant(s) as a prize.

© 1996 Critical Thinking Books & Software • P.O. Box 448, Pacific Grove, CA 93950 • 800-458-4849

16 Pick 'Em Up!

Challenge

Invent a contraption that will pick up walnuts.

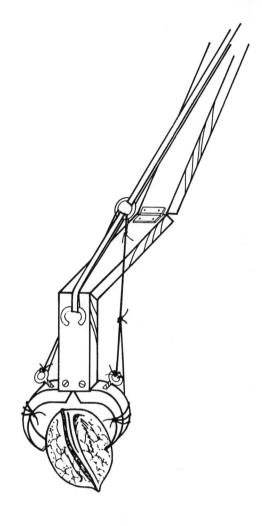

Rules

1. The teacher will supply the walnuts, but you must supply everything else needed to build your contraption.

2. No more than 5 different parts may be used in the construction of your contraption.

3. Things used to attach or secure parts together—screws, wire, nails, nuts, bolts, string, tape, etc. are NOT counted as parts.

4. No electrical energy may be used.

5. You may not bend over when using your contraption. You must remain upright at all times.

6. Your walnut contraption must be built within the time limit set by the teacher.

7. The contraption that picks up the most walnuts in the shortest period of time will be declared the winning design.

16   For the Teacher

The Challenge

This nutty activity challenges students to invent a device that might help people with a bad back pick up objects off the floor.

The Rules

1. Have students work alone or in teams.

2. You will need to provide walnuts or similar objects to be picked up. I use walnuts because they are easy to get and last indefinitely, but other objects will work. Something smaller, like marbles, would be even more challenging.

3. Make sure students understand that they must build a device from scratch to meet this challenge. Commercial products should not be allowed.

4. You can have students either work on their designs during class or on their own outside of class. I present this challenge to my students then give them a reasonable amount of time to work on their designs outside of class. The competition is held during class time. If your students work in teams, one class period should be sufficient to allow everyone to test their designs. To save time, urge students to have their designs fully built and tested before the day of the competition.

5. Set a time limit and inform students of what this time limit will be before they begin constructing their devices.

6. Consider setting this competition up in a tournament format. Match students or teams against each other and put them in brackets. In the first round, set a time limit of several minutes. The winners of the first round advance to the second round. Make the time limit somewhat longer for the second round. Eventually you will come down to two students or teams in the final or championship round. Make the final round the longest time limit. If you are not familiar with setting up tournament brackets, enlist the aid of your coaches or P.E. staff. They can also show you how to go about determining the order of finish beyond first and second place.

7. Inform students that they will be allowed to adjust and/or repair their devices between rounds only if they can do so in a reasonable amount of time.

8. You will need a stopwatch for timing purposes and a large open area in which to conduct the competition. Pushing the desks back in a classroom will usually provide enough space.

Safety Tips

If students work on their designs outside of class, you may not have a clue about the safety of the materials and components they are using. Therefore, I suggest you require students to submit a diagram of their design for your approval before they begin construction.

Awards and Recognition

1. If you go with a tournament format, display the brackets and results where they can be appreciated by other students, teachers, and parents.

2. Display the actual devices each student or team constructed.

3. Consider giving the winning student or team a bag of nuts as a prize.

4. You could paint a walnut an appropriate color, glue it to a small piece of wood, and present it to the winning student or team as a trophy.

17 Radical Racers

Challenge

Design and build a vehicle that will travel as far as possible using only the energy from a mousetrap.

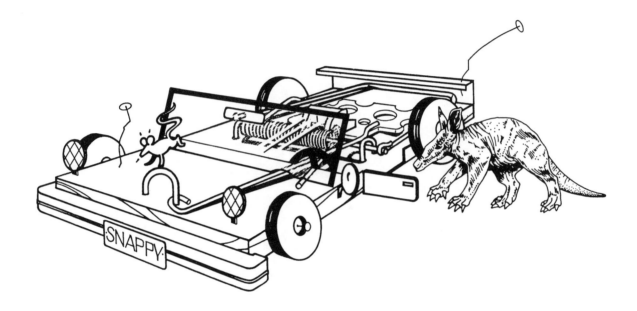

Rules

1. The teacher will supply you with one mousetrap. You must supply everything else you need.

2. The mousetrap may be altered in any way you wish.

3. The mousetrap may be attached to other parts or other parts may be attached to the mousetrap.

4. The spring of the mousetrap must be the only energy source used to move the contraption.

5. The contraption must move on its own. You cannot push or help your contraption in anyway.

6. Your car must be built within the time limit set by the teacher.

7. If time permits, you will be given several trial runs. Only the longest run will be counted.

8. You will be allowed to repair or adjust your contraption between trial runs.

9. The contraption that travels the greatest distance will be declared the winning design.

17 For the Teacher

The Challenge

In this activity, students are challenged to turn a mousetrap into a car that can travel as far as possible using only the energy from the spring of the mousetrap.

The Rules

1. Have students work alone or in teams. From a time and materials standpoint, I find it more convenient to have students work in teams of two or three.

2. You will need to supply each student or team with one mousetrap. Common mousetraps can be reasonably purchased at hardware or grocery stores.

3. You can have students work on their designs either during class or outside of class. I present this challenge to my students, hand them their mousetraps, and then give them several days to build their devices on their own. The devices are brought to school for the day of the competition. One class period is usually sufficient to give each team several trial runs.

4. These cars tend to go all over the place and some of them may travel surprisingly far, so I suggest you hold the competition in a large open area like a gymnasium. You will also need a measuring tape to determine the distance each car travels. A 20-foot tape will suffice, but a 50-foot tape works the best.

5. Place a one- to two-foot piece of masking tape on the floor in an appropri-ate spot and use a marker to draw a dark line the length of the tape. This will be the starting line. At the start of each run, students must align their cars so that the front ends of their cars are on the starting line.

6. On your command, have students release their cars and stand back. No one may push or otherwise aid any car.

7. Once each car has traveled its maxi-mum, measure the distance traveled from the starting line to the front end of the car.

8. Give students or teams as many trial runs as time permits. Count only the longest run for each car.

9. Allow students a reasonable amount of time between trial runs to adjust or repair their cars as needed.

10. You can make this activity more diffi-cult by varying the challenge to see which car can come the closest to trav-eling a specific distance. Mark start and stop lines on the floor a reason-able distance apart—10 to 15 feet— and challenge students to design their cars so that they stop as close as pos-sible to the stop line. Points could be awarded. The closer to the stop line, the higher the points. The car with the most points at the end of the trial runs would be the winning design.

Safety Tips

Caution students to keep their fingers away from set mousetraps.

18 See and Do

Challenge

Turn accurate observations into usable instructions.

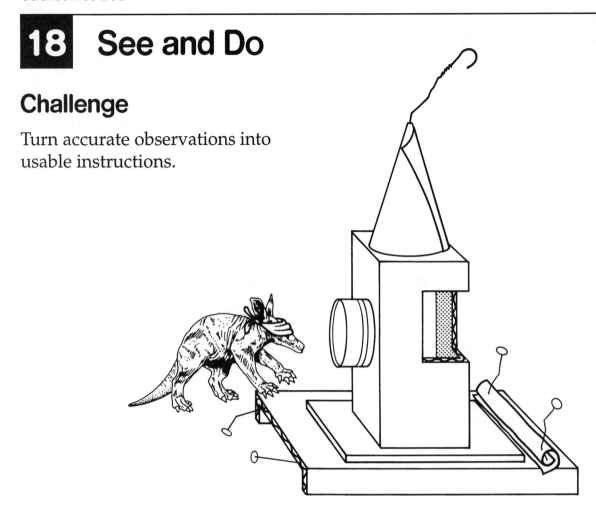

Rules

1. You will work as a team with another student.

2. One team member will be the observer/writer and the other team member will be the doer.

3. The teacher will show a contraption to only the observer/writer. The observer/writer will be allowed to carefully examine and observe the contraption.

4. The observer/writer must then write a set of instructions for building a contraption exactly like the one built by the teacher. No diagrams or pictures are allowed. The contraption will then be taken apart.

5. The doer must now take the observer/writer's instructions and build a contraption exactly like the one built by the teacher using the parts from the teacher's original contraption.

6. A perfect reconstruction of the teacher's original contraption is worth 50 points. Each mistake made in reconstruction will result in a 5-point deduction.

7. The teacher will then build another contraption and you will switch roles.

8. The team with the highest combined point total will be crowned as the champion observers/writers/doers.

18 For the Teacher

The Challenge

Everyone has problems following instructions, but in this activity students will also come to appreciate how hard it is to write usable instructions.

The Rules

1. Have students work alone in teams of two. Designate one student as the observer/writer and the other student as the doer.

2. Show only the observer/writer a contraption you have constructed. Tinkertoys or Lego blocks work well for this. The observer/writer should not watch you construct the contraption but see it for the first time only in its final state.

3. Give the observer/writer adequate time to observe and investigate your contraption. The observer/writer should then write a set of instructions for building such a contraption as he/she disassembles your contraption.

4. Once the instructions are written and your contraption has been disassembled, bring in the doer. The doer must use the instructions and the disassembled pieces to reconstruct your original contraption.

5. You should keep a separate display copy or picture of each contraption so that students can compare their reconstruction with the original.

6. Determine the points earned.

7. Build another contraption or have another contraption already built and allow the students to switch roles.

8. Again determine the points earned. Total all points earned. The team with the highest combined point total is the champion observer/writer/doer.

9. A variation of this activity would be to have the observer/writer read the instructions to the doer. Thus, you could challenge students to follow written and/or vocal instructions.

Safety Tips

Use common sense when building and dismantling contraptions.